42 WONDERS OF OUR NIGHT SKY

Damonscotting.com
@DamonScotting
@DamonScotting
@ThisIsAstronomical

First published 10th March 2026
First edition

ISBN (paperback): 9781919563206
ISBN (hardback): 9781919563213

Independently published

Image credits and third-party acknowledgements are listed in the "Acknowledgements and Image Credits" section of this book.

Astronomy safety notice: Never look at the Sun with the naked eye, binoculars, or a telescope without a properly fitted solar filter designed for that instrument. Permanent eye damage can occur instantly.

This book is dedicated to the teachers who helped me find my path.

Thank you to Guy & Steve for giving your time and teaching me how to teach.

Thank you to Randip for teaching me to focus on engaging kids with wonder.

Thank you to Ann for helping spark my interest in astronomy.

And thank you to Mum and Dad for always reminding me to get a real job.
I hope you're happy now! ❤

Contents

ABOUT THE AUTHOR

Damon Scotting is an award-winning astrophotographer, published writer, and qualified science teacher with a talent for making space science feel simple. Through his online astronomy education, he's helped a global audience learn what's up there, and how to see it for themselves.

In The 42 Wonders of Our Night Sky, Damon rounds up the most jaw-dropping sights in the cosmos and shows you what they really look like through everyday backyard setups and state-of-the-art space telescopes.

2. ANDROMEDA GALAXY 86
3. ORION NEBULA 85
PLEIADES 84
5. SATURN 83
6. JUPITER 81
8. MOON 80
9. ALBIREO 80
10. SUN 79
SOMBRERO GALAXY 79
12. WHIRLPOOL GALAXY 79
13. HELIX NEBULA 78
14. FIREWORKS GALAXY 78
16. ELEPHANT TRUNK 77
17.
BETELGEUSE 75
20. MIZAR & ALCOR 74
21. HORSEHEAD NEBULA 74
HINDS V. NEBULA 73
24. MARS 73
25. M87 72
26. MESSIER 3
29. HERCULES CLUSTER 70
30. LEO TRIPLET 69
31. HEART NEBULA
M81 & M82 66
34. RING NEBULA 66
35. WHALE GALAXY 65
36. THORS HELM
VENUS 61
39. ALGOL 60
40. BEEHIVE CLUSTER 51
DOUGIE
42 WONDERS NIGHT SKY

WONDERWALL
42 WONDERS NIGHT SKY

AMELIA
DOUGIE
42 WONDERS NIGHT SKY
42 WONDERS OF OUR NIGHT SKY

How to use the 42 WONDERS OF OUR NIGHT SKY Official BOOK

The 42 Wonders of Our Night Sky began as a YouTube series, inspired by the "mission-style" storytelling of Steve Backshall's Deadly 60: find something incredible, learn what it is, and rank it. With one key message at the end: now go find this wonder for yourself!

This book is a guide to 42 of the most amazing objects in our night sky. In each entry you'll learn what the Wonder is, how to find it from your garden, and what to expect through three very different setups:

$500: Seestar smart telescope
$1,000,000: Telescope Live remote observatories
$1,000,000,000: Space telescopes (Hubble & James Webb)

Each Wonder is rated for Beauty, Power, and Mystery. The average of those three scores becomes its Wonder Rating, which determines its position on the *Wonderwall.*

You'll also see a Visibility Rating (out of 5) based on how easy it is to observe from your garden, plus a How to Spot It chart with key stats, the best time to look, and where to find it.

IMPORTANT

Scan the QR code at the bottom-right of every page to jump to the exact moment I talk about that Wonder in the official YouTube series.

Stargazing is one of the easiest ways to reconnect with something bigger than your day-to-day and I hope this book gives you the push to step outside, look up, and fall in love with the night sky.

What You can See

The same Wonder captured through 3 VERY different setups

Seestar S50

$500

Remote observatory

$1,000,000

Space Telescope

$1,000,000,000

Beauty	Power	Mystery	Wonder Rating
??	??	??	??

Visibility rating: ★★☆☆☆

HOW TO SPOT IT

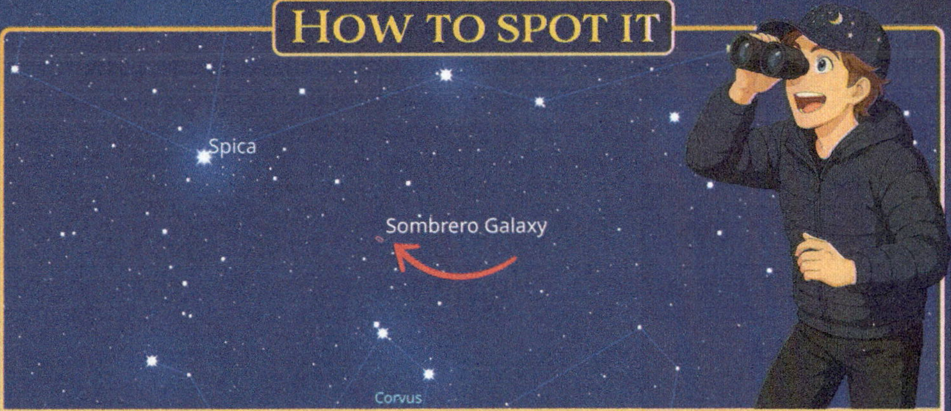

- Magnitude (Brightness): 8
- Visible to the naked eye: ☒
- Distance: 29,000,000 Light years
- Object catalog name: M104

Have you spotted it yet? ☐

- Constellation: Virgo
- Visibility rating: ★★☆☆☆
- Best time to see it: February - May
- Coordinates: 12h 39m 59s, -11° 37' 23"

TELESCOPES

Seestar S50

$500

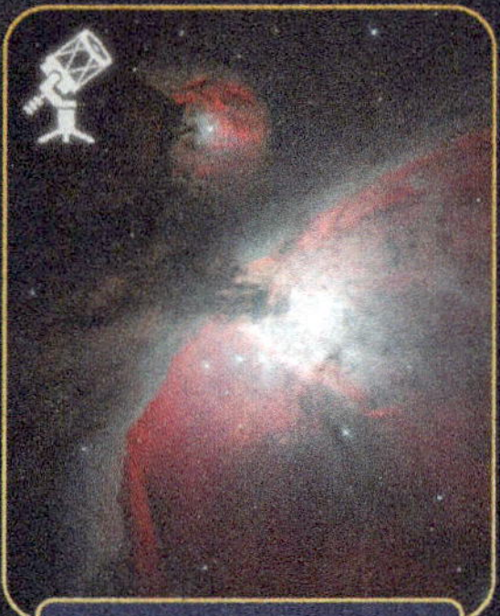

Remote observatory

$1,000,000

Space Telescope

$1,000,000,000

Seestar S30/S50

Seestar smart telescopes have changed astrophotography forever by making it genuinely accessible. They're an all-in-one mini observatory that requires no complicated setup, no messing around with lots of fancy equipment.

Just tap a target in the app and you're imaging deep-sky objects in minutes, with built-in info as you go. I'm a huge fan of the smart telescopes I'd have loved to have had one when I started.

Throughout this series I use each of the following telescopes:

Seestar S30 ($349)

Seestar S50 ($499)

Seestar S30 Pro ($599)

Telescope Live gives you remote access to a network of telescopes under dark skies, so you can capture deep-sky data without owning any equipment.

They offer 100,000s of affordable images to download and edit yourself as well as tutorials to learn from.

Plans start at $6.

Remote observatory

The Hubble Space Telescope has produced some of the most iconic space images in history. While its discoveries will live on, Hubble is expected to be retired in the 2030s.

When the mission ends, NASA intends to send the space telescope out in a blaze of glory via a controlled re-entry into the atmosphere.

In which the telescope will burn up with any remaining debris targeted to fall harmlessly into a remote stretch of ocean. Fortunately, we have the superior powers of the James Webb Telescope to see us into the next age of exploration.

Space Telescope

WONDERS FEATURED

Episode 1

Episode 2

Episode 3

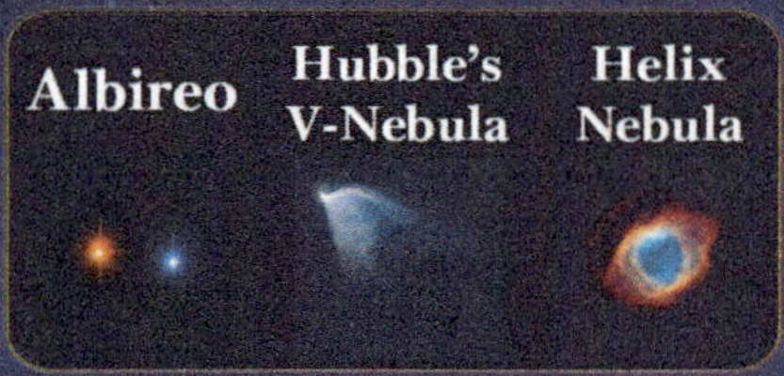

Episode 4

Episode 5

Episode 6

Episode 7

WONDERS FEATURED

Episode 8

Ring Nebula | Whirlpool Galaxy | Uranus

Episode 9

Algol | Moon | Devil in Disguise

Episode 10

Betelgeuse | Fireworks Galaxy | Cigar Galaxy

Episode 11

Pleiades | Venus | Elephant Trunk Nebula

Episode 12

Mizar & Alcor | N America Nebula | Whale Galaxy

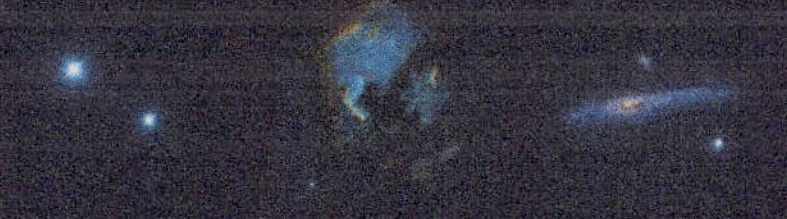

Episode 13

Pillars of Creation | Needle Galaxy | Rosette Nebula

Episode 14

Veil Nebula | Pinwheel Galaxy | Crab Nebula

Orion Nebula

Ranked: 3rd

Visibility rating: ★★★★★

Beauty	Power	Mystery	Wonder Rating
95	72	89	85

Why it is in the list?

You are witnessing the emphatic birthplace of new stars. The Orion Nebula is a stellar nursery of the highest order. Visible as a faint blurry patch of light to the naked eye it is a legend amongst astrophotographers.

← 1,344 Light years away →

Visible just below the belt of Orion in its sword is a treasure trove of cosmic beauty. You could make a very strong case for the Orion nebula being the ultimate Wonder of our Night Sky. For many astrophotographers it was their first ever deep sky object.

What You can See

The same Wonder captured through 3 VERY different setups

Seestar S50

$500

Remote observatory

$1,000,000

Space Telescope

$1,000,000,000

FUN FACT

Whereas a lot of the wonders in this series are best observed by those living in the northern hemisphere, the orion nebula sits along the celestial equator line in our night sky, meaning it is a favourite amongst stargazers across the globe. It is one of the most accessible deep sky objects out there!

HOW TO SPOT IT

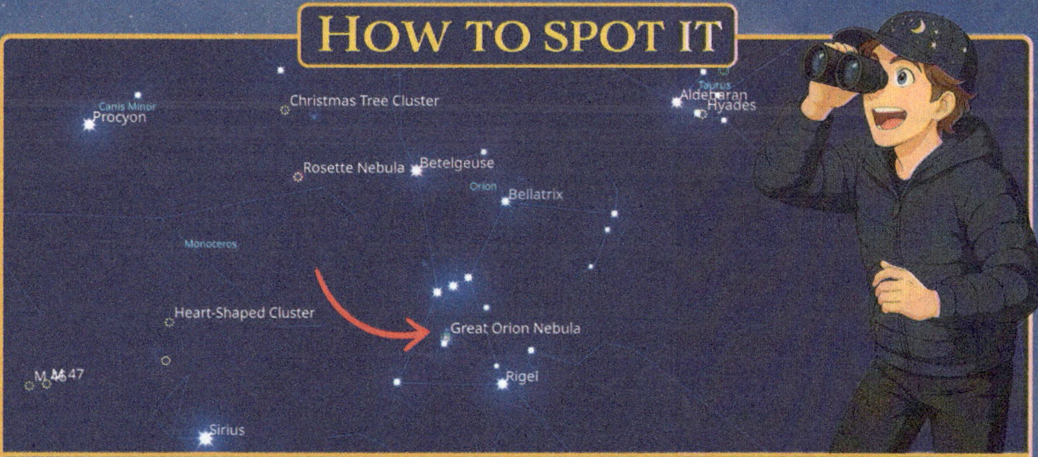

- Magnitude (Brightness): 4
- Visible to the naked eye: ☑
- Distance: 1,344 Light years
- Object catalog name: M42/NGC 1976

Have you spotted it yet? ☐

- Constellation: Orion
- Visibility rating: ★★★★★
- Best time to see it: December - Feb
- Coordinates: 05h 35m 17s, -05° 23' 28"

Hercules Globular Cluster

Ranked: 29th

Beauty	Power	Mystery	Wonder Rating
84	78	49	70

Why it is in the list?

The great globular cluster in Hercules is one of the brightest and best-known globular clusters in the Northern Hemisphere.

← 22,000 Light years away →

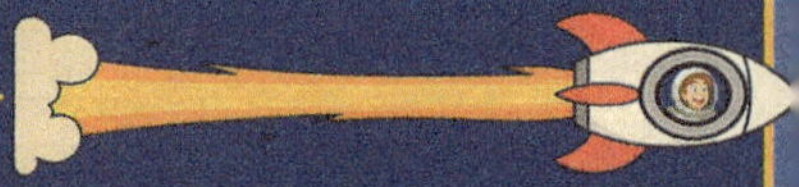

In the central region, stars are so densely packed they can be just 0.1 light-years apart, much closer than our Sun's nearest neighbour. Some theories suggest globular clusters like M13 are the remnants of dwarf galaxies consumed by the Milky Way long ago.

What You can See

The same Wonder captured through 3 VERY different setups

Seestar S50

$500

Remote observatory

$1,000,000

Space Telescope

$1,000,000,000

FUN FACT

Hello Aliens? In 1974, astronomers sent a powerful radio transmission from the Arecibo Observatory, toward the globular star cluster M13. Because M13 is about 22,000 light-years away, the signal would take roughly 22,000 years to arrive.

HOW TO SPOT IT

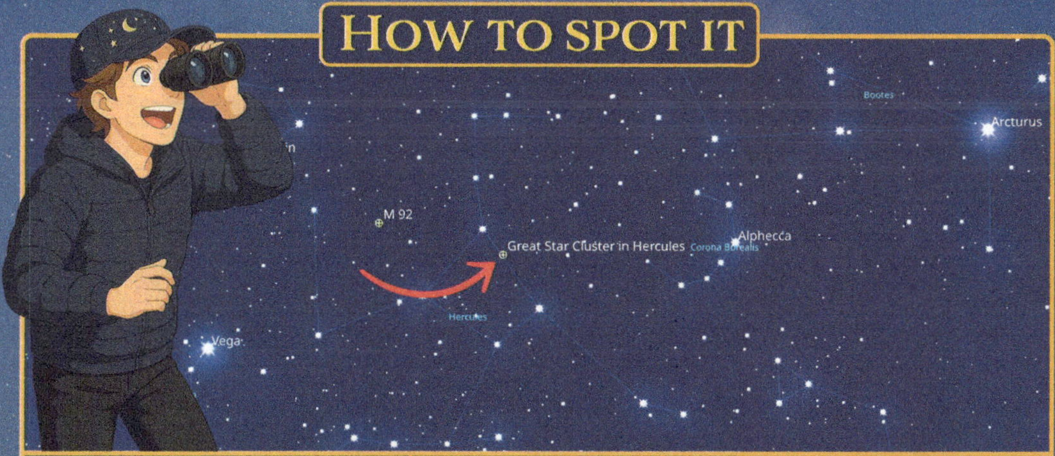

- Magnitude (Brightness): 5.8
- Visible to the naked eye: ☑
- Distance: 22,000 Light years
- Object catalog name: M13

Have you spotted it yet? ☐

- Constellation: Hercules
- Visibility rating: ★★★☆☆
- Best time to see it: June-August
- Coordinates: 16h 41m 41s, +36° 27′ 35″

Sombrero Galaxy

Ranked: 11th

Beauty	Power	Mystery	Wonder Rating
84	79	74	79

Why it is in the list?

The stars are mostly old and evolved indicating this is a mature galaxy and likely 13.25 billions of years old. Making it almost as old as the universe itself.

← 29,000,000 Light years away →

This galaxy is classified as a lenticular galaxy which means it's actually a mix between a spiral and an elliptical galaxy

It has over 2,000 globular clusters (the Milky Way has 150-200 making this an unusually large number).

What You can See

The same Wonder captured through 3 VERY different setups

Seestar S50

$500

Remote observatory

$1,000,000

Space Telescope

$1,000,000,000

FUN FACT

The galaxy gets its name because of its unique shape: a bright central bulge and a flat disc surrounded by a prominent dust lane resembling a sombrero hat.

HOW TO SPOT IT

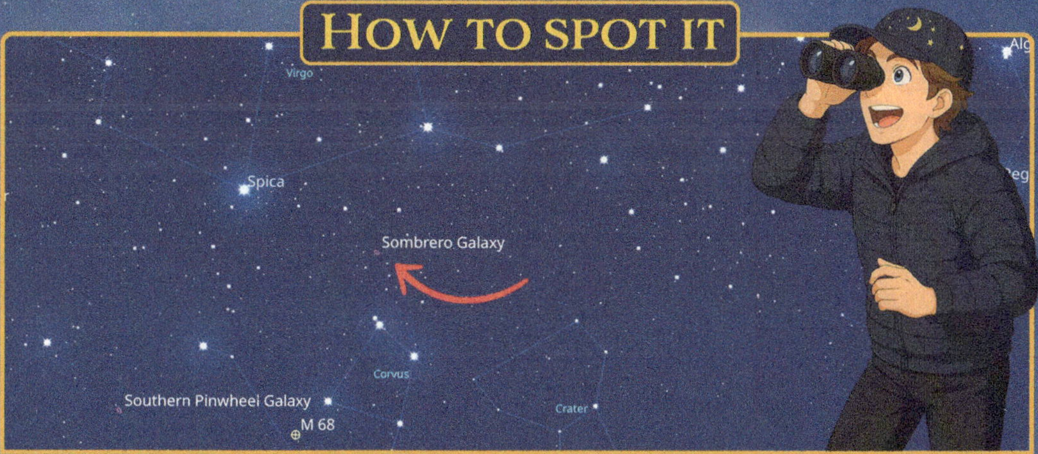

- Magnitude (Brightness): 8
- Visible to the naked eye: ☒
- Distance: 29,000,000 Light years
- Object catalog name: M104

Have you spotted it yet? ☐

- Constellation: Virgo
- Visibility rating: ★★☆☆☆
- Best time to see it: February - May
- Coordinates: 12h 39m 59s, -11° 37′ 23″

SATURN

Ranked: 5th

Visibility rating: ★★★★☆

Beauty	Power	Mystery	Wonder Rating
97	73	78	83

WHY IT IS IN THE LIST?

If any single object has turned more people into stargazers, it's this one.

← 1,200,000,000 Kilometres away →

Its famous rings are younger than the dinosaurs and may vanish within 100 million years, they are a fleeting spectacle on cosmic timescales.

Saturn is the only planet in our solar system that is less dense than water, meaning if you could find a large enough body of water it would float!

What You can See

The same Wonder captured through 3 VERY different setups

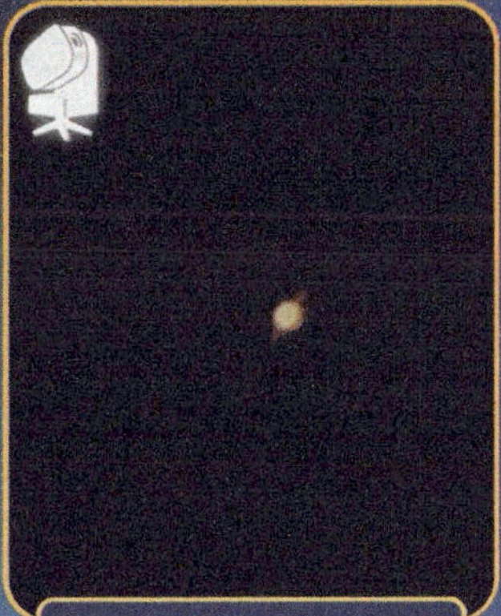

Seestar S50

$500

Remote observatory

$1,000,000

Space Telescope

$1,000,000,000

FUN FACT

Saturn has at least 274 confirmed moons, the most of any planet by far.
It is practically its own mini solar system!

HOW TO SPOT IT

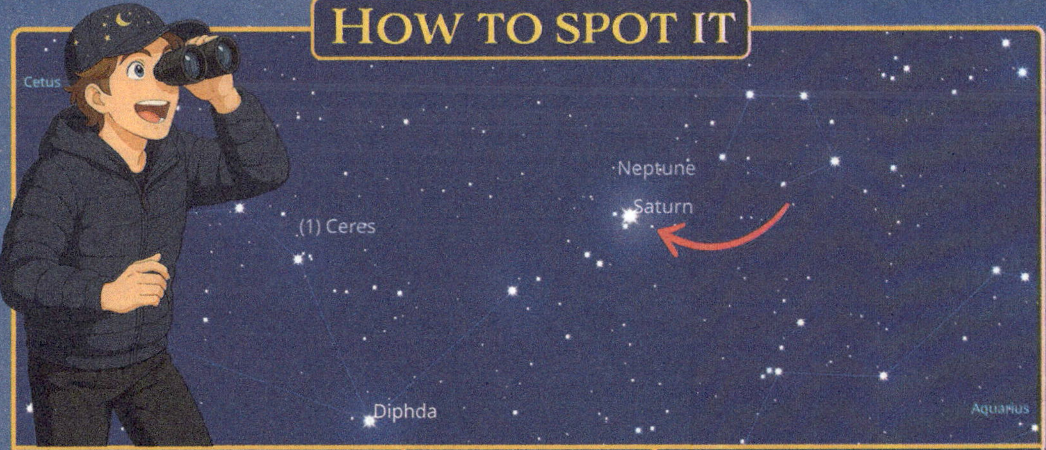

- Magnitude (Brightness): 0.46 (peak)
- Visible to the naked eye: ☑
- Distance: 1,200,000,000 km
- Object catalog name: Saturn

Have you spotted it yet? ☐

- Constellation: Varies
- Visibility rating: ★★★★☆
- Best time to see it: Every 12.4 Months
- Coordinates: Varies

Heart Nebula

Ranked: 31st

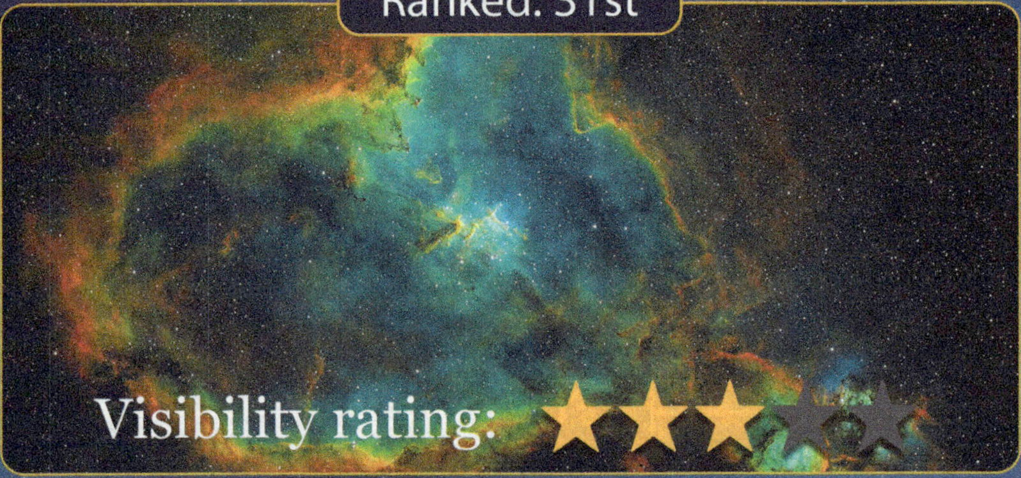

Visibility rating: ★★★☆☆

Beauty	Power	Mystery	Wonder Rating
96	67	40	68

Why it is in the list?

At the center of the nebula is a cluster of young, hot stars called Melotte 15, blasting out radiation and stellar winds that carve the nebula's swirling, arched heart shape. It's as if the nebula is alive and beating.

← 7,500 Light years away →

The Heart Nebula is a perfect example of beauty hidden in plain sight.

It's invisible to the naked eye, but once its delicate structure is revealed, it's unforgettable.

What You can See

The same Wonder captured through 3 VERY different setups

Seestar S50

$500

Remote observatory

$1,000,000

Space Telescope

$1,000,000,000

FUN FACT

The nebula can be observed throughout most of the year by those who live in the northern hemisphere but the best time to see it high in your skies is around Valentine's day in February.

HOW TO SPOT IT

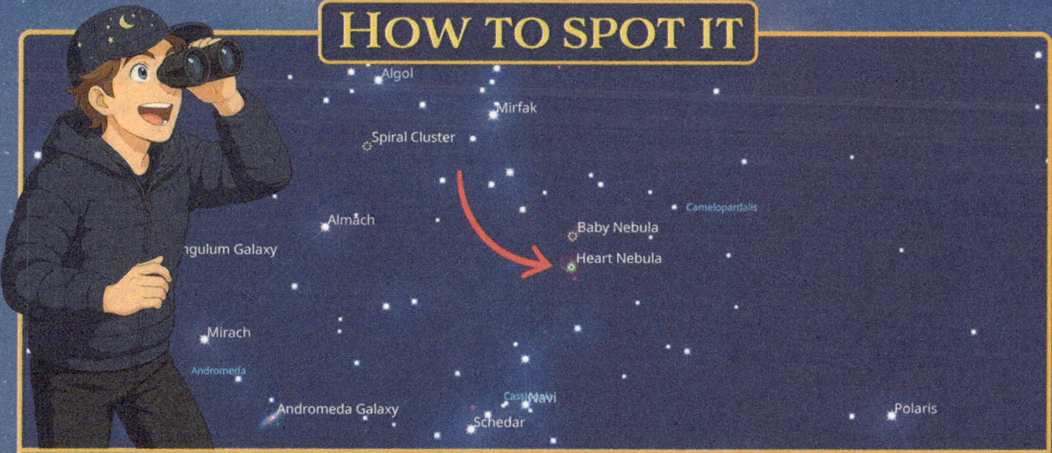

- Magnitude (Brightness): 6.5
- Visible to the naked eye: ☒
- Distance: 7,500 Light years
- Object catalog name: IC 1805

Have you spotted it yet? ☐

- Constellation: Cassiopeia
- Visibility rating: ★★★☆☆
- Best time to see it: October-April
- Coordinates: 02h 33m 22s, +61° 26′ 36″

A Supermassive Black Hole

Ranked: 25th

Visibility rating: ★☆☆☆☆

Beauty	Power	Mystery	Wonder Rating
17	100	100	72

Why it is in the list?

Messier object 87 is a giant elliptical galaxy, no spiral arms, just a massive ball of stars and one of the largest galaxies in the nearby universe.
It is a freak of nature...

← 54,000,000 Light years away →

M87 is famous for its powerful jet of plasma, which extends for thousands of light years from its central supermassive black hole. It sits alongside a whole host of galaxies in Markarian's Chain.

What You can See

The same Wonder captured through 3 VERY different setups

Seestar S50

$500

Remote observatory

$1,000,000

Space Telescope

$1,000,000,000

FUN FACT

In 2019, M87 became world-famous when the Event Horizon Telescope (EHT) released the first image of a black hole's shadow.

HOW TO SPOT IT

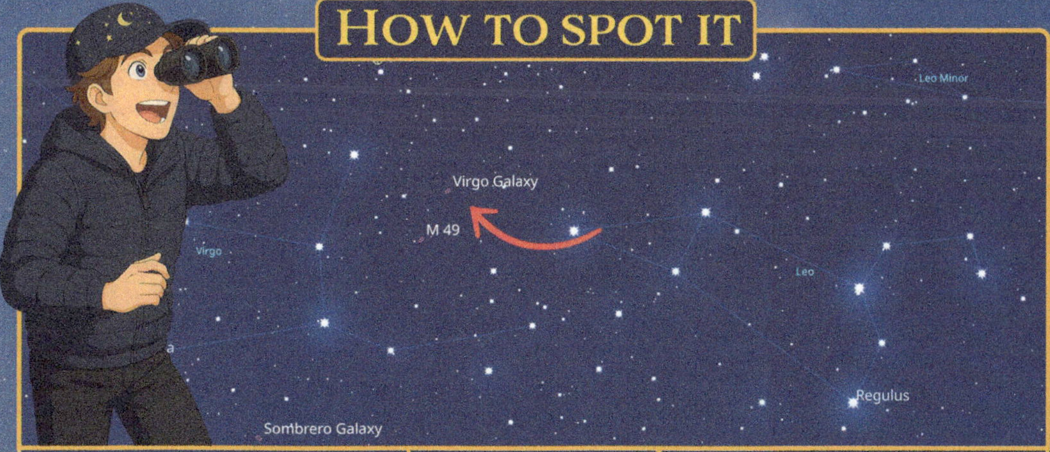

- Magnitude (Brightness): 8.6
- Visible to the naked eye: ☒
- Distance: 54,000,000 Light years
- Object catalog name: M87/NGC 4486

Have you spotted it yet? ☐

- Constellation: Virgo
- Visibility rating: ★☆☆☆☆
- Best time to see it: March - May
- Coordinates: 12h 30m 49s, +12° 23' 28"

The Sombrero Galaxy (11th) & Saturn (5th) as captured by the Hubble Space Telescope.

HELIX NEBULA

Ranked: 13th

Visibility rating: ★★☆☆☆

Beauty	Power	Mystery	Wonder Rating
96	61	78	78

WHY IT IS IN THE LIST?

This planetary nebula is often known as the 'Eye of God' since it is famous for its striking resemblance to a giant eye. It looks like someone is watching us from space...

← 655 Light years away →

One day, in 5–6 billion years, our Sun will do exactly this: shed its outermost layers and leave behind a lonely white dwarf surrounded by glowing gas so technically you are looking at the future ghost of our solar system.

What You can See

The same Wonder captured through 3 VERY different setups

Seestar S50

$500

Remote observatory

$1,000,000

Space Telescope

$1,000,000,000

FUN FACT

The gas is flying outward at 32 km/s, meaning this ghostly cloud is still growing as you read this.

HOW TO SPOT IT

- Magnitude (Brightness): 7.6
- Visible to the naked eye: ☒
- Distance: 655 Light years
- Object catalog name: NGC 7293

Have you spotted it yet? ☐

- Constellation: Aquarius
- Visibility rating: ★★☆☆☆
- Best time to see it: June-October
- Coordinates: 22h 29m 39s, -20° 50' 14"

HUBBLE'S VARIABLE NEBULA

Ranked: 22nd

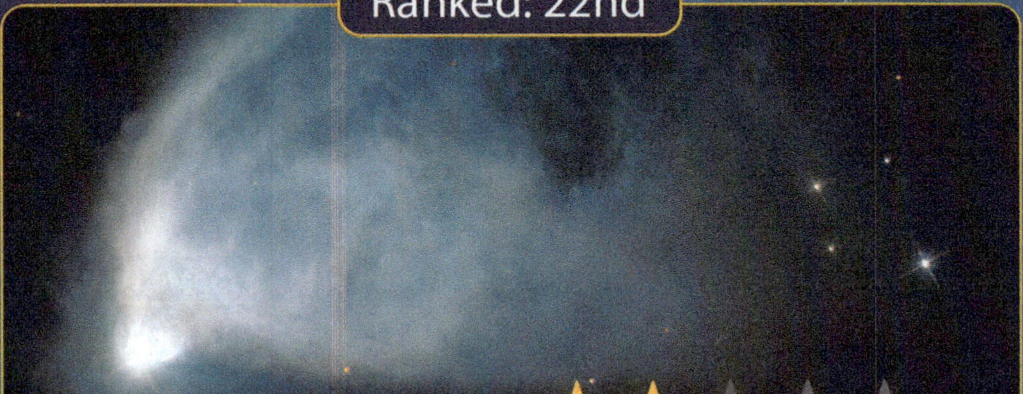

Visibility rating: ★★☆☆☆

Beauty	Power	Mystery	Wonder Rating
63	70	86	73

WHY IT IS IN THE LIST?

Unlike most nebulae, which evolve over thousands of years, Hubble's Variable Nebula can visibly change in just days or weeks. Which means no two photographs of it will look the same!

← 2,500 Light years away →

The nebula has a bright "head" near the star and a fanned-out, cone-like "tail," making it look a lot like a comet frozen in space. Imagine the frustration of early-day comet hunters trying to spot the latest new comet appearance and constantly seeing a sight like this variable nebula!

What You can See

The same Wonder captured through 3 VERY different setups

Seestar S50

$500

Remote observatory

$1,000,000

Space Telescope

$1,000,000,000

FUN FACT

The nebula's appearance is caused by dense dust clouds near the star R Monocerotis, casting shadows outwards leading to these ominous light ripples.

HOW TO SPOT IT

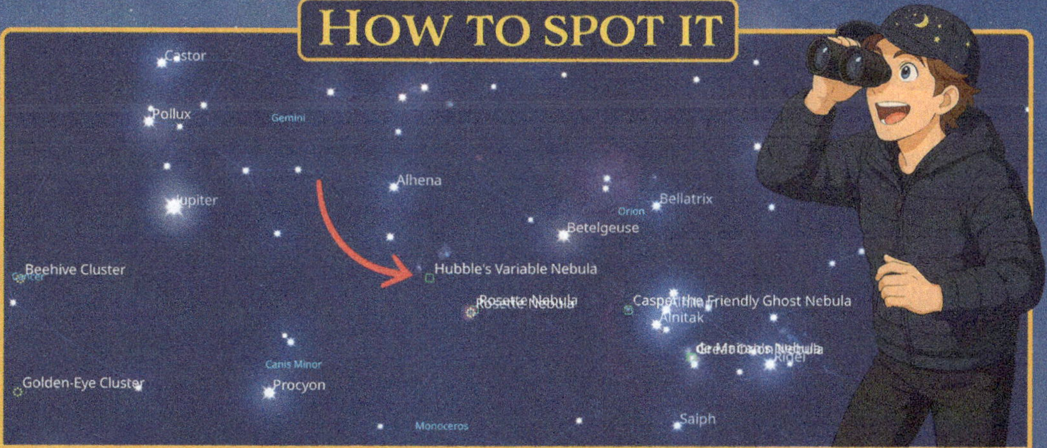

- Magnitude (Brightness): 9
- Visible to the naked eye: ☒
- Distance: 2,500 Light years
- Object catalog name: NGC 2261

Have you spotted it yet? ☐

- Constellation: Monoceros
- Visibility rating: ★★☆☆☆
- Best time to see it: September-May
- Coordinates: 6h 39m 10s, +08° 45′ 00″

ALBIREO

Ranked: 9th

Visibility rating: ★★★★☆

Beauty	Power	Mystery	Wonder Rating
96	66	77	80

WHY IT IS IN THE LIST?

Albireo is arguably the most beautiful binary star in our night sky. The two stars are different colours due to temperature difference.

← 433.8 Light years away →

Albireo A (orange) is a cool giant star, 4,000 K while Albireo B (blue) is a hotter star, 13,000 K. The pair is an easy split in a small telescope: their separation is about 35 arcseconds, so even modest magnification shows two distinct stars.

What You can See

The same Wonder captured through 3 VERY different setups

Seestar S50

$500

Remote observatory

$1,000,000

My 16" Telescope

$2,000

FUN FACT

While appearing as a single point of light to the naked eye, a telescope reveals two distinct stars: a bright orange-yellow star (Albireo A) and a fainter blue-green star (Albireo B).

HOW TO SPOT IT

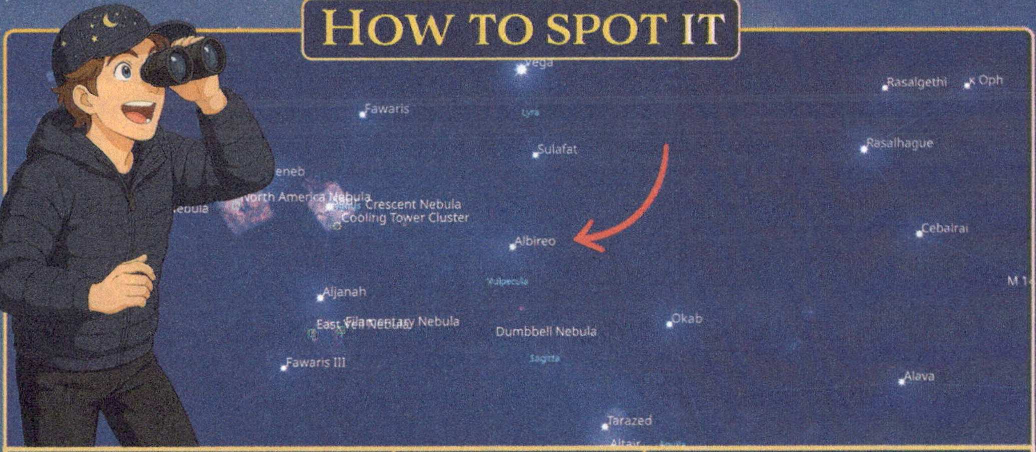

- Magnitude (Brightness): 3.1
- Visible to the naked eye: ☑
- Distance: 434 Light years
- Object catalog name: β Cygni

Have you spotted it yet? ☐

- Constellation: Cygnus
- Visibility rating: ★★★★☆
- Best time to see it: August-September
- Coordinates: 19h 30m 43s, +27° 57' 34"

JUPITER

Ranked: 6th

Visibility rating: ★★★★★

Beauty	Power	Mystery	Wonder Rating
76	87	81	81

WHY IT IS IN THE LIST?

Jupiter's massive gravity acts like a cosmic vacuum cleaner, pulling in or deflecting many asteroids and comets. Without it, Earth might experience far more catastrophic impacts.

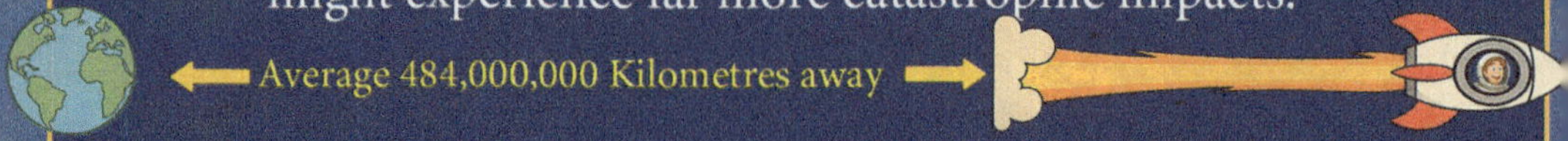

Jupiter is a gas giant which means it doesn't have any solid surfaces like Earth so if you tried to land on it you would sink into layers of gas and liquid with crushing pressures!

Jupiter's famous Great Red Spot is a storm larger than Earth and it's been raging for at least 350 years! Winds reach up to 432 km/hr.

What You can See

The same Wonder captured through 3 VERY different setups

Seestar S50

$500

Remote observatory

$1,000,000

Space Telescope

$1,000,000,000

FUN FACT

Jupiter is so big you could fit 1,300 Earths inside it! But despite being the largest planet, Jupiter has the shortest day in our solar system lasting just 10 hours.

HOW TO SPOT IT

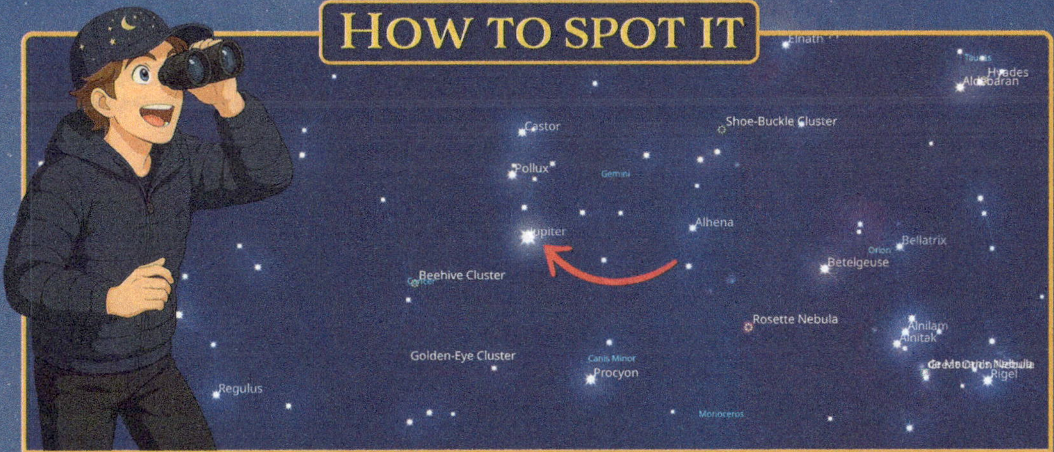

- Magnitude (Brightness): -2.94 (peak)
- Visible to the naked eye: ☑
- Distance: 484,000,000 Km away
- Object catalog name: Jupiter

Have you spotted it yet? ☐

- Constellation: Varies
- Visibility rating: ★★★★★
- Best time to see it: Every 13 Months
- Coordinates: Varies

MESSIER 3

Ranked: 26th

Visibility rating: ★★★☆☆

Beauty	Power	Mystery	Wonder Rating
80	70	66	72

WHY IT IS IN THE LIST?

This immense star cluster is over 11 billion years old making it almost as old as the universe itself. It is one of the biggest and brightest star clusters visible in our night sky.

← 33,920 Light Years away →

M3 is one of the richest sources for a specific type of variable star. By imaging the star cluster over the course of the night with a sensitive camera you can document how it gently pulses over the time span of even a few hours.

What You can See

The same Wonder captured through 3 VERY different setups

Seestar S50

$500

Remote observatory

$1,000,000

Space Telescope

$1,000,000,000

FUN FACT

M3 is an ancient relic of the early universe, likely formed when the Milky Way itself was still a baby.
If you lived on a planet orbiting one of the stars at the centre your nights would be permanently illuminated by starlight.

HOW TO SPOT IT

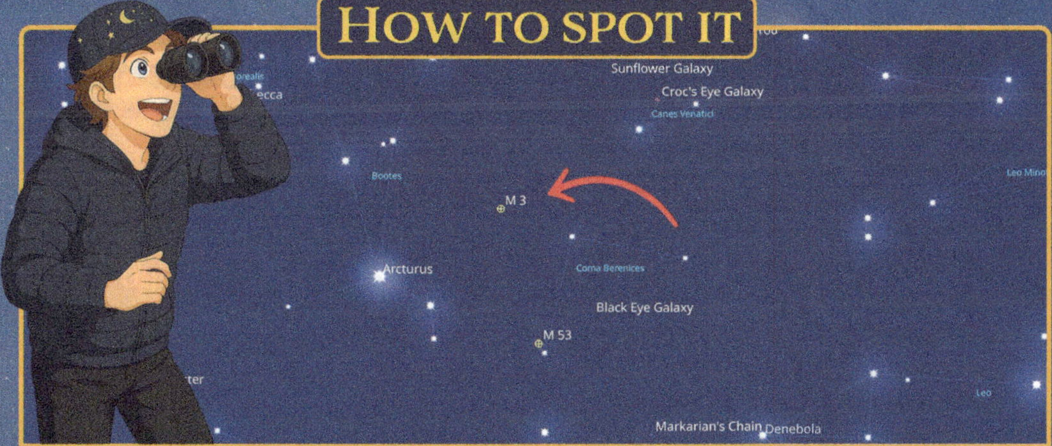

- Magnitude (Brightness): 6.2
- Visible to the naked eye: ☒
- Distance: 33,920 Light Years
- Object catalog name: M3/NGC 5272

Have you spotted it yet? ☐

- Constellation: Canes Venatici
- Visibility rating: ★★★☆☆
- Best time to see it: March-June
- Coordinates: 13h 42m 12s, +28° 22′ 38″

Leo Triplet

Ranked: 30th

Visibility rating: ★☆☆☆☆

Beauty	Power	Mystery	Wonder Rating
64	86	58	69

Why it is in the list?

The Leo Triplet consists of three interacting spiral galaxies located about 35 million light-years away. They are commonly known as M65, M66 and NGC 3628 (Hamburger Galaxy).

← 35,000,000 Light Years away →

Tidal forces between the galaxies are actively distorting them, especially NGC 3628, which shows a huge dark dust lane and tidal tail stretching over 300,000 light-years.

M66 is a starburst galaxy, with enhanced star formation triggered by gravitational interactions with its neighbours.

What You can See

The same Wonder captured through 3 VERY different setups

Seestar S50

$500

Remote observatory

$1,000,000

Space Telescope

$1,000,000,000

FUN FACT

The three galaxies (M65, M66, and NGC 3628) are gravitationally bound, meaning they are constantly pulling on each other, causing their shapes to distort.

HOW TO SPOT IT

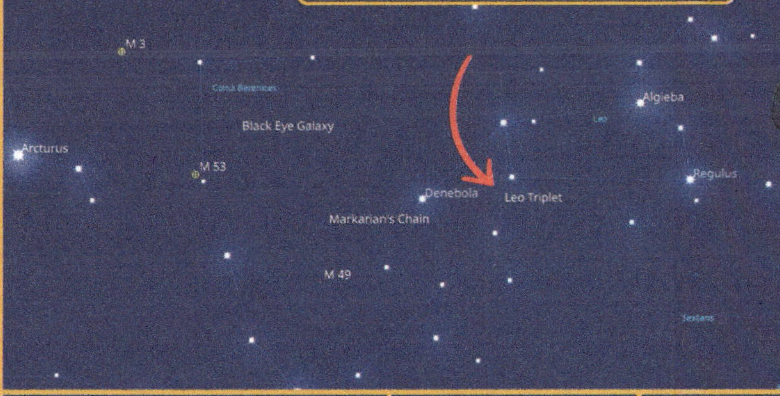

- Magnitude (Brightness): 9.7
- Visible to the naked eye: ☒
- Distance: 35,000,000 Light Years
- Object catalog name: M65, M66, NGC 3628

Have you spotted it yet? ☐

- Constellation: Leo
- Visibility rating: ★☆☆☆☆
- Best time to see it: April-May
- Coordinates: 11h 19m 49s, +13° 13' 28"

DUMBBELL NEBULA

Ranked: 37th

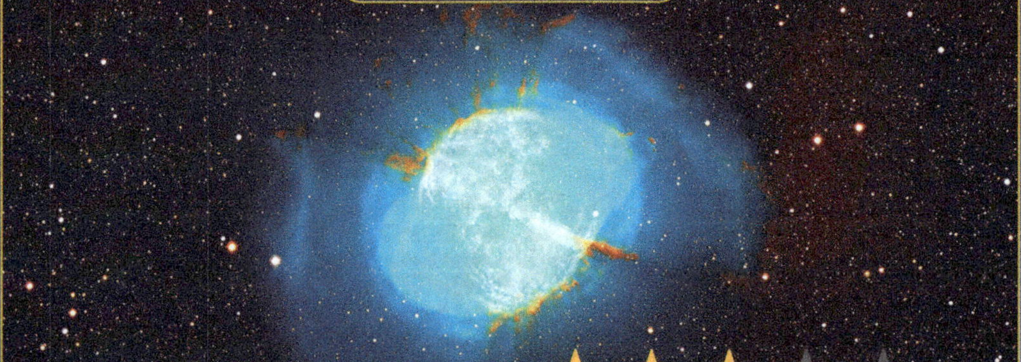

Visibility rating: ★★★☆☆

Beauty	Power	Mystery	Wonder Rating
71	61	55	62

WHY IT IS IN THE LIST?

The Dumbbell nebula, also known as the apple core nebula, is famously the first planetary nebula ever to be discovered, first spotted by Charles Messier in 1764.

← 1,360 Light Years Away →

It is one of the brightest planetary nebula in the sky, making it one of the easiest of its kind to observe with small telescopes and even binoculars under dark skies.
It is a fantastic target for beginner astrophotographers thanks to its bright colourful appearance.

What You can See

The same Wonder captured through 3 VERY different setups

Seestar S50

$500

Remote observatory

$1,000,000

Space Telescope

$1,000,000,000

FUN FACT

The Dumbbell nebula spans about 2.5 light-years across and is slowly expanding into space at a speed of roughly 30 km/s!

HOW TO SPOT IT

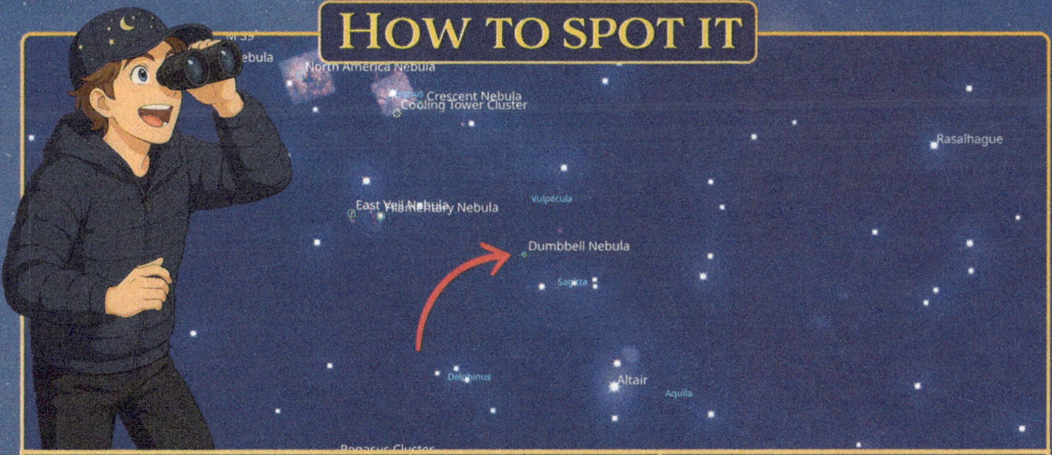

- Magnitude (Brightness): 7.5
- Visible to the naked eye: ☒
- Distance: 1,360 Light Years Away
- Object catalog name: M27/NGC 6853

Have you spotted it yet? ☐

- Constellation: Vulpecula
- Visibility rating: ★★★☆☆
- Best time to see it: July-September
- Coordinates: 19h 59m 36s, +22° 43′ 16″

INTERNATIONAL SPACE STATION

Ranked: 41st

Visibility rating: ★★★★★

Beauty	Power	Mystery	Wonder Rating
53	19	77	50

WHY IT IS IN THE LIST?

The ISS is the only man-made wonder on our list, weighing in at about 420,000 kg, which is about the same as 320 cars.

← 400 kilometres away →

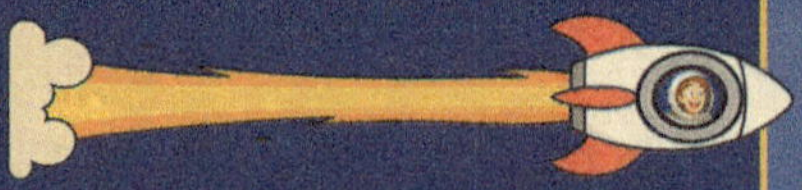

The ISS orbits Earth at a speed of 17,500 mph, it completes a full circle around the planet every 90 minutes.
This means that the astronauts aboard the International Space Station get to see 16 sunrises and 16 sunsets every single day!

What You can See

The Wonder captured through my Celestron 8" Edge HD

Celestron 8" SCT - $2000

Fun Fact

Astronauts have to work out for two hours a day while aboard to help keep their muscles in shape. With no gravity compressing their spines, astronauts can grow in height by up to 2 inches within their first couple of days aboard the ISS.

How to spot it

- Magnitude (Brightness): -5.9 (Peak)
- Visible to the naked eye: ☑
- Distance: 400 km
- Object catalog name: ISS

Have you spotted it yet? ☐

- Constellation: Varies
- Visibility rating: ★★★★★
- Best time to see it: Varies
- Coordinates: Varies

ANDROMEDA GALAXY

Ranked: 2nd

Visibility rating: ★★★★☆

Beauty	Power	Mystery	Wonder Rating
78	80	99	86

WHY IT IS IN THE LIST?

Andromeda is the galaxy next door, when it was first identified as a galaxy it expanded our horizons further than ever before by confirming that our galaxy was not the limit of our universe.

← 2,500,000 Light Years away →

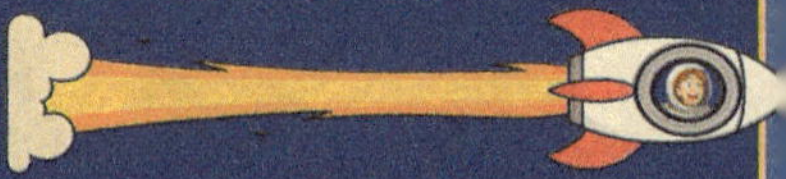

The Andromeda Galaxy is visible to the naked eye without the aid of a telescope, in fact it's widely regarded as the most distant object us humans can see with just our eyes. From dark sky sites the galaxy appears as a tiny faint cloudy patch of light, the truth is, it's an island containing trillions of stars.

What You can See

The same Wonder captured through 3 VERY different setups

Seestar S50

$500

Remote observatory

$1,000,000

Space Telescope

$1,000,000,000

FUN FACT

Andromeda is on a collision course with the Milky Way, and they are predicted to merge in about 4.5 billion years. Forming a new galaxy titled "Milkdromeda".

HOW TO SPOT IT

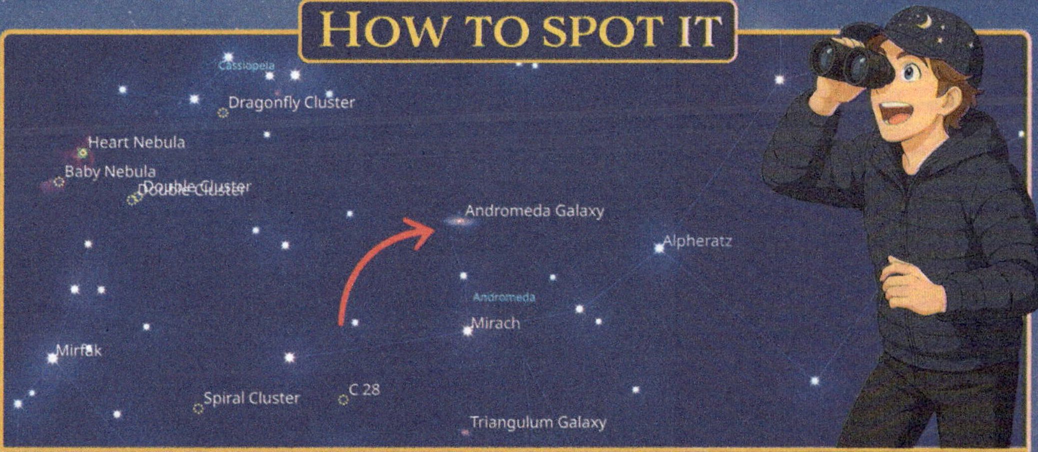

- Magnitude (Brightness): 3.4
- Visible to the naked eye: ☑
- Distance: 2,500,000 Light Years
- Object catalog name: M31/NGC 224

Have you spotted it yet? ☐

- Constellation: Andromeda
- Visibility rating: ★★★★☆
- Best time to see it: September-November
- Coordinates: 00h 42m 44s, +41° 16' 09"

HORSEHEAD NEBULA

Ranked: 21st

Beauty	Power	Mystery	Wonder Rating
85	68	69	74

WHY IT IS IN THE LIST?

The Horsehead Nebula's shape is not permanent. Over tens of thousands of years, stellar winds and radiation will slowly erode and reshape the Horsehead.

← 1,375 Light Years Away →

The red background glow is from hydrogen gas ionized by nearby bright stars like Sigma Orionis. Without that glow, the Horsehead would be invisible against the black of space!
A cosmic silhouette roughly 4 light years in height.

What You can See

The same Wonder captured through 3 VERY different setups

Seestar S50

$500

Remote observatory

$1,000,000

Space Telescope

$1,000,000,000

Fun Fact

The infrared view from the Hubble Space Telescope reveals the secrets hidden within the Horsehead. Stars otherwise shielded by the thick clouds of dust and gas.

How to spot it

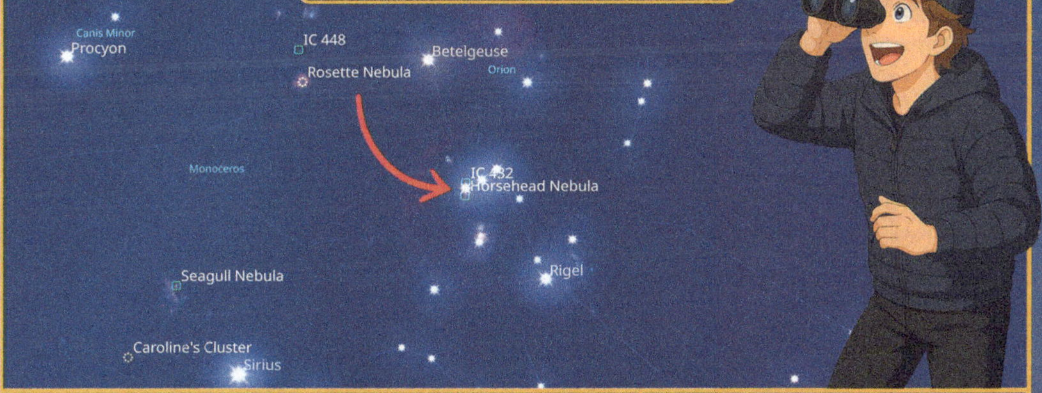

- Magnitude (Brightness): 7.4
- Visible to the naked eye: ☒
- Distance: 1,375 Light Years Away
- Object catalog name: B33

Have you spotted it yet? ☐

- Constellation: Orion
- Visibility rating: ★★★☆☆
- Best time to see it: December-February
- Coordinates: 05h 40m 59s, -02° 27' 30"

MARS

Ranked: 24th

Visibility rating: ★★★★★

Beauty	Power	Mystery	Wonder Rating
91	32	95	73

WHY IT IS IN THE LIST?

Mars has frozen water and carbon dioxide at its poles which grow in the winter and shrink in the summer. You can actually see these with your own amateur telescope.

← Average 227,800,000 kilometres Away →

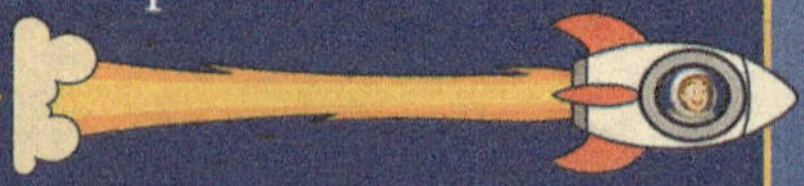

It gets its reddish colour from iron oxide (rust) on the surface. The gravity on Mars is just 38% of that here on Earth, can you imagine how fun it would be to experience weighing a third of your current weight!

What You can See

The same Wonder captured through 3 VERY different setups

Seestar S50

$500

Remote observatory

$1,000,000

Space Telescope

$1,000,000,000

FUN FACT

Mars is home to the largest volcano in the solar system, called Olympus Mons. This giant is around three times the height of Mount Everest and covers an area equivalent to that of France.

HOW TO SPOT IT

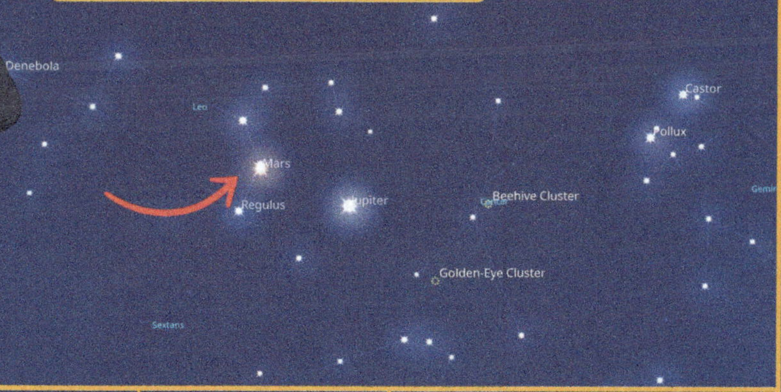

- Magnitude (Brightness): -2.9 (peak)
- Visible to the naked eye: ☑
- Distance: Average 227,800,000 km Away
- Object catalog name: Mars

Have you spotted it yet? ☐

- Constellation: Varies
- Visibility rating: ★★★★★
- Best time to see it: Every 26 months
- Coordinates: Varies

ANTENNAE GALAXY

Ranked: 18th

Visibility rating: ★★★★☆

Beauty	Power	Mystery	Wonder Rating
97	88	44	76

WHY IT IS IN THE LIST?

The Antennae Galaxies are actually two spiral galaxies crashing into each other. They're in the process of merging, creating one giant elliptical galaxy over hundreds of millions of years.

← 50,000,000 Light Years Away →

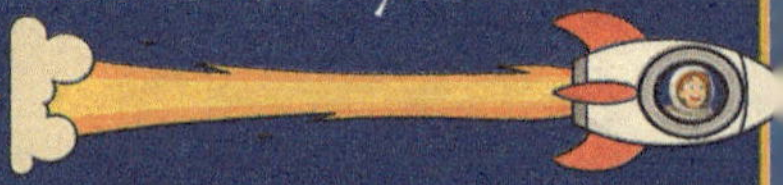

The Antennae Galaxies show us what might happen when our Milky Way and the Andromeda Galaxy collide in about 4–5 billion years. But I doubt that our "Milkdromeda" merger will look as beautiful as the Antennae galaxy!

What You can See

The same Wonder captured through 3 VERY different setups

Seestar S50

$500

Remote observatory

$1,000,000

Space Telescope

$1,000,000,000

FUN FACT

They are named for the long, antenna-like tidal tails that extend from them due to the ongoing collision. This cosmic dance of galaxies is a prime example of galactic mergers and is a rich source of new stars.

HOW TO SPOT IT

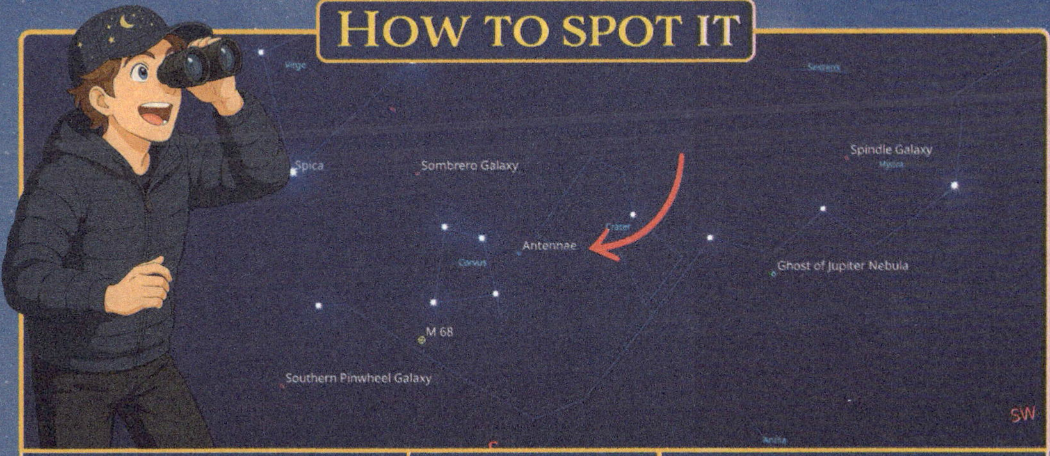

- Magnitude (Brightness): 11.2
- Visible to the naked eye: ☒
- Distance: 50,000,000 Light Years Away
- Object catalog name: NGC 4038/4039

Have you spotted it yet? ☐

- Constellation: Corvus
- Visibility rating: ★☆☆☆☆
- Best time to see it: April - June
- Coordinates: 12h 01m 53s, -18° 52' 10"

Orion's Belt & Sword (right) contain 2 Wonders, the Horsehead nebula (12th) and Orion nebula (3rd). Its next door neighbour the Rosette nebula (27th) (Bottom left) can be seen amongst a bright colourful starfield. Captured by the $599 Seestar S30 Pro Smart Telescope.

THOR'S HELMET

Ranked: 36th

Visibility rating: ★☆☆☆☆

Beauty	Power	Mystery	Wonder Rating
74	65	49	63

WHY IT IS IN THE LIST?

The star inside Thor's Helmet is a Wolf-Rayet star which means it is in a late evolutionary phase, and it may go supernova any day...

← 11,960 Light Years away →

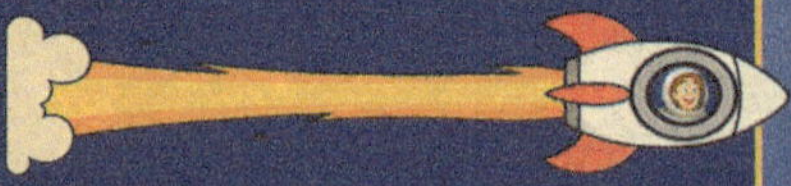

Before going supernova it begins shedding its most outer layers. The helmet-like structure was formed by superfast winds from the central star slamming into the surrounding gas. These interactions create shock waves, shaping the nebula.

What You can See

The same Wonder captured through 3 VERY different setups

Seestar S50

$500

Remote observatory

$1,000,000

Space Telescope

$1,000,000,000

FUN FACT

The central star (WR 7) is a rare, ultra-hot, ultra-massive star that's shedding its outer layers at a ridiculous rate. It's not often we spot stars in a pre-supernova phase but Thor's Helmet is already putting on a pretty spectacular show for us.

HOW TO SPOT IT

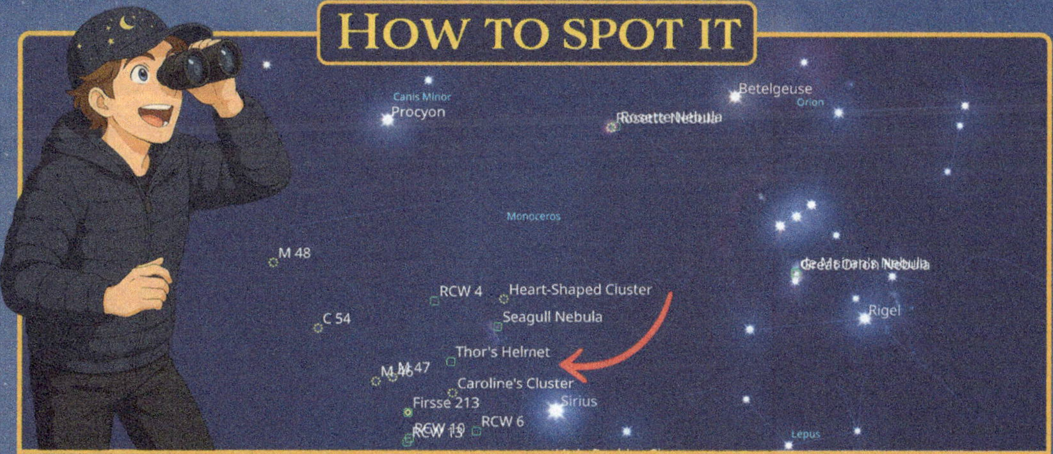

- Magnitude (Brightness): 11.45
- Visible to the naked eye: ☒
- Distance: 11,960 Light Years
- Object catalog name: NGC 2359

Have you spotted it yet? ☐

Scan to watch more!

- Constellation: Canis Major
- Visibility rating: ★☆☆☆☆
- Best time to see it: October-May
- Coordinates: 07h 18m 30s, -13° 13′ 48″

BEEHIVE CLUSTER

Ranked: 40th

Beauty	Power	Mystery	Wonder Rating
71	34	49	51

WHY IT IS IN THE LIST?

This open star cluster is easily visible to the naked eye and caught the attention of early astronomers as they thought it looked like a swarm of bees buzzing around, hence the name!

← 594 Light years Away →
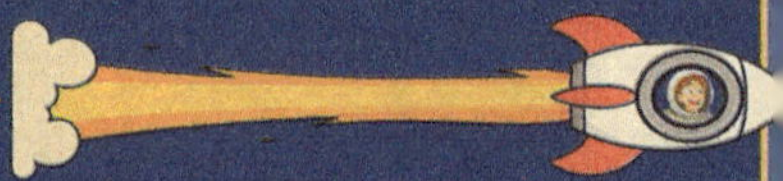

The stars in the Beehive move together through space, like a cosmic family, which is how we know they were born together. Even when viewed from a heavily light polluted garden with just a small pair of binoculars you can still easily see this bright cluster of stars.

What You can See

The same Wonder captured through 3 VERY different setups

Seestar S50

$500

Remote observatory

$1,000,000

Space Telescope

$1,000,000,000

FUN FACT

Because the Hubble Space Telescope's field of view is so narrow it actually looks well beyond the star cluster, revealing faint galaxies that are otherwise not picked up with other telescopes!

HOW TO SPOT IT

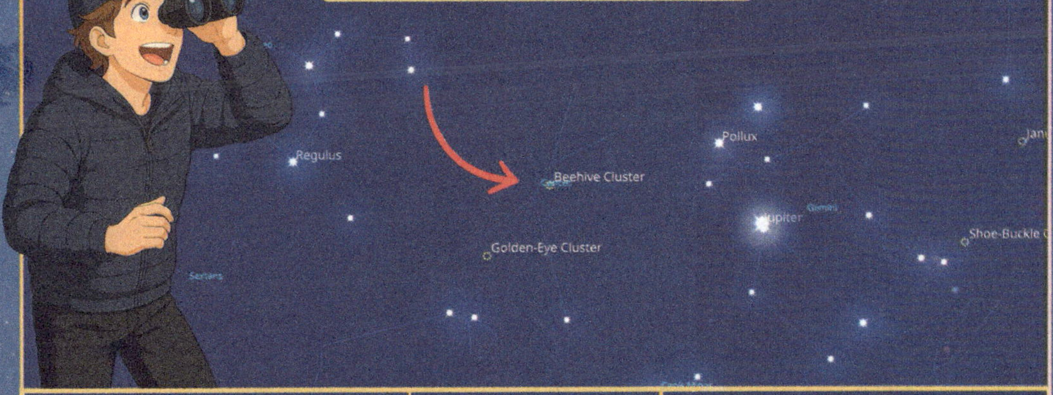

- Magnitude (Brightness): 3.7
- Visible to the naked eye: ☑
- Distance: 594 Light Years Away
- Object catalog name: M44/NGC 2632

Have you spotted it yet? ☐

- Constellation: Cancer
- Visibility rating: ★★★★☆
- Best time to see it: February-May
- Coordinates: 08h 40m 00s, +19° 59′ 00″

SUN

Ranked: 10th

Visibility rating: ★★★★★

Beauty	Power	Mystery	Wonder Rating
65	91	82	79

WHY IT IS IN THE LIST?

The Sun provides nearly all the energy for life on Earth. You could make a very good case for it being number 1 on our list, just like you could also argue it shouldn't even be on the list at all.

← 150,000,000 Kilometres Away →

Our Sun goes through an 11-year solar cycle, switching from calm to stormy, in 2025 it reached its most recent peak leading to aurora storms visible across the planet. At around 4.6 billion years old, the Sun is roughly halfway through its life.

What You can See

The same Wonder captured through 3 VERY different setups

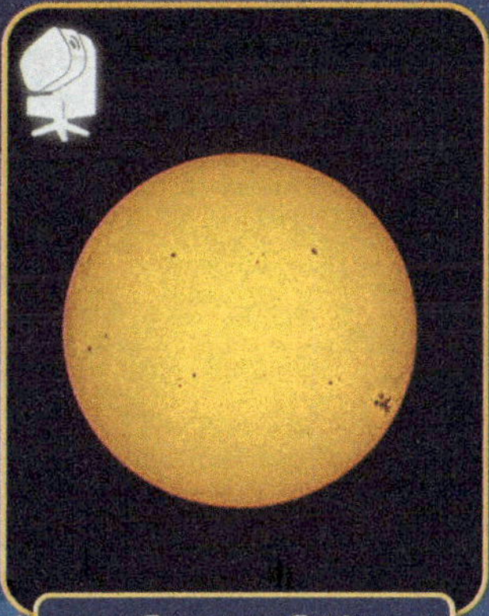

Seestar S50

$500

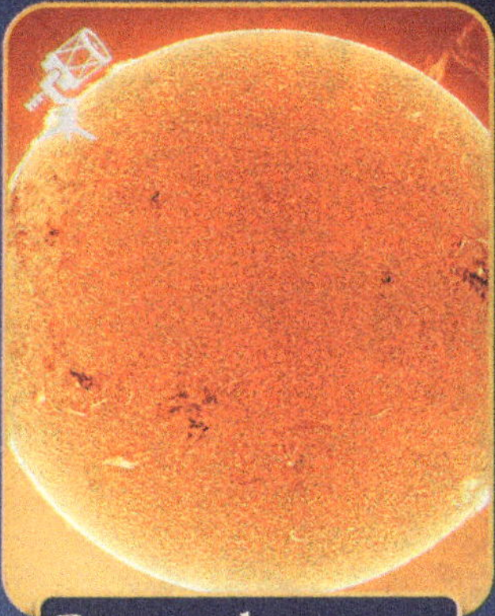

Remote observatory

$1,000,000

Space Telescope

$1,000,000,000

Fun Fact

In another 5-6 billion years, it will become a red giant, then shrink to a white dwarf. Whilst it is the largest object in our solar system, there are trillions of stars in our universe many of which are larger than it...

How to spot it

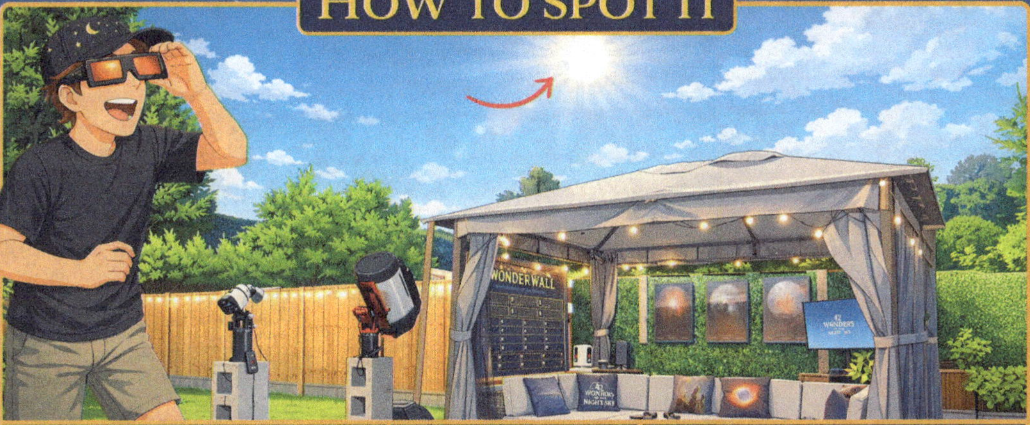

- Magnitude (Brightness): -26.74
- Visible to the naked eye: ☑
- Distance: 150,000,000 km Away
- Object catalog name: Sol

Have you spotted it yet? ☐

- Constellation: N/A
- Visibility rating: ★★★★★
- Best time to see it: Everyday (Does not apply to British)
- Coordinates: N/A

WHIRLPOOL GALAXY

Ranked: 12th

Beauty	Power	Mystery	Wonder Rating
90	94	52	79

WHY IT IS IN THE LIST?

Does it get any more EPIC than two galaxies colliding with one another?! The Whirlpool galaxy (M51) is being penetrated by its smaller companion, NGC 5194.

← 31,000,000 Light years away →

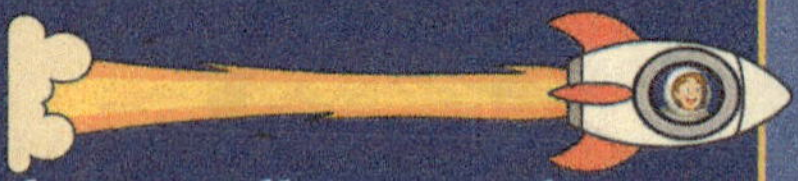

Its dramatic spiral structure and ongoing galactic collision make it a favourite target for backyard astronomers eager to capture one of the finest examples of galactic interaction.
The Whirpool galaxy was actually the first galaxy ever to be specifically classified as a spiral galaxy!

What You can See

The same Wonder captured through 3 VERY different setups

Seestar S50

$500

Remote observatory

$1,000,000

Space Telescope

$1,000,000,000

FUN FACT

Its companion, NGC 5195, is not peacefully orbiting, but plunging through the Whirlpool, triggering intense waves of new star birth!

HOW TO SPOT IT

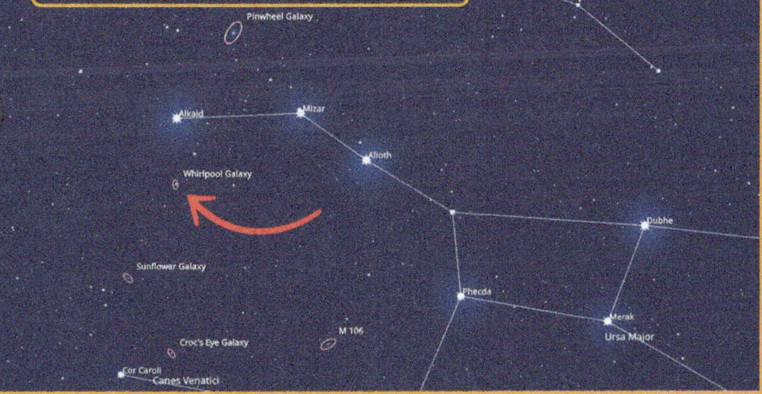

- Magnitude (Brightness): 8.4
- Visible to the naked eye: ☒
- Distance: 24,000,000 Light years
- Object catalog name: M51/NGC 5194

Have you spotted it yet? ☐

- Constellation: Canes Venatici
- Visibility rating: ★★★☆☆
- Best time to see it: February - May
- Coordinates: 13h 29m 53s, +47° 11' 43"

Ring Nebula

Ranked: 34th

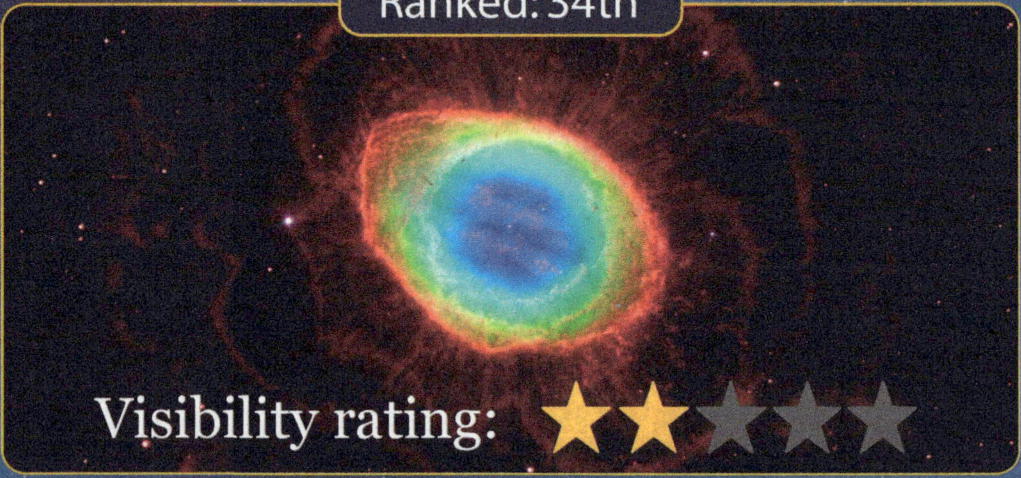

Visibility rating: ★★☆☆☆

Beauty	Power	Mystery	Wonder Rating
67	59	71	66

Why it is in the list?

The Ring Nebula formed when a star similar to the Sun ran out of fuel and shed its most outer layers forming what is known as a planetary nebula.

← 2,000 Light Years Away →

The colourful light comes from ionized gases; red from hydrogen and green/blue from oxygen.

PRO TIP:

Through a telescope the Ring Nebula is best seen by looking off centre. This method of averted vision helps pick up its faint details.

What You can See

The same Wonder captured through 3 VERY different setups

Seestar S50

$500

Remote observatory

$1,000,000

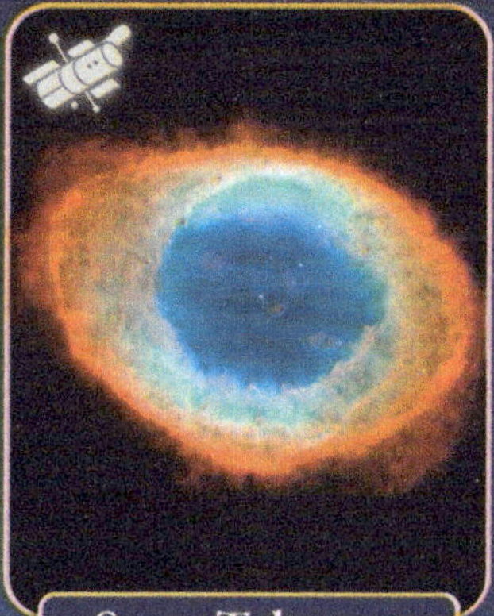

Space Telescope

$1,000,000,000

Fun Fact

At the Ring Nebula's centre is a white dwarf, the exposed core of a once Sun-like star. This tiny remnant is blasting intense ultraviolet radiation into the surrounding gas, causing it to glow beautifully. A small glimpse of what our Sun may look like in about 6 billion years' time.

How to spot it

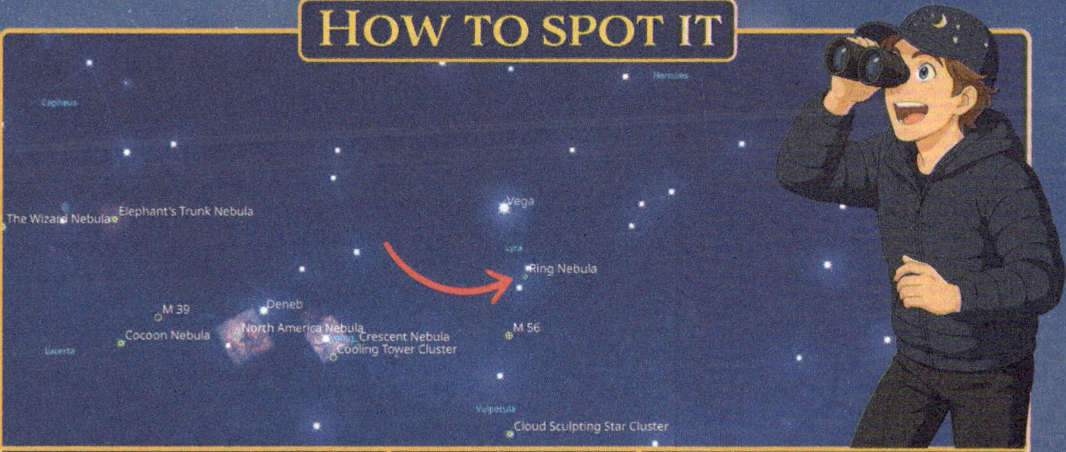

- Magnitude (Brightness): 8.8
- Visible to the naked eye: ☒
- Distance: 2,000 Light Years Away
- Object catalog name: M57/NGC 6720

Have you spotted it yet? ☐

- Constellation: Lyra
- Visibility rating: ★★☆☆☆
- Best time to see it: July-August
- Coordinates: 18h 53m 35s, +33° 01′ 45″

URANUS

Ranked: 42nd

Visibility rating: ★★★☆☆

Beauty	Power	Mystery	Wonder Rating
39	61	48	49

WHY IT IS IN THE LIST?

Uranus is a gas giant, which whilst not sounding too appealing, has a beautiful appearance due to the methane gas in its atmosphere, it absorbs red light and reflects blue.

← 1,900,000,000 Kilometres away →

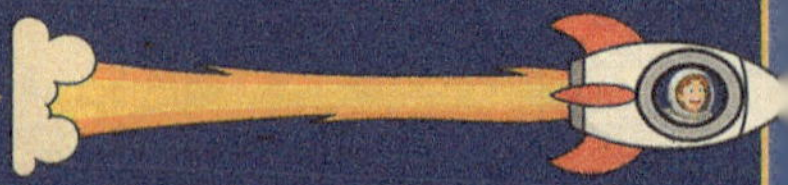

Uranus is a very cold and windy world. The ice giant is surrounded by 13 faint rings and 28 small moons. The seventh planet from the Sun rotates at a 97.8-degree angle from the plane of its orbit. This unique tilt makes Uranus appear to spin sideways, orbiting the Sun like a rolling ball.

What You can See

The same Wonder captured through 3 VERY different setups

Seestar S50

$500

Remote observatory

$1,000,000

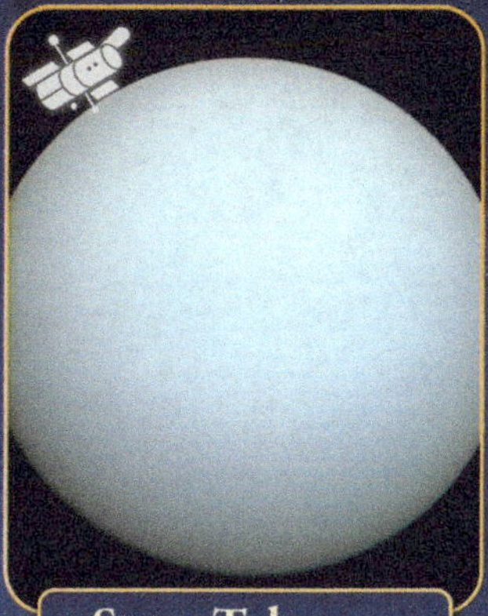

Space Telescope

$1,000,000,000

Fun Fact

Because Uranus is tilted on its side by 98 degrees, each pole experiences 42 years of continuous daylight followed by 42 years of darkness, giving the planet the most extreme seasons in the Solar System...

How to spot it

- Magnitude (Brightness): 5.7
- Visible to the naked eye: ☑
- Distance: 1,900,000,000 Km
- Object catalog name: Uranus

Have you spotted it yet? ☐

- Constellation: Varies
- Visibility rating: ★★★☆☆
- Best time to see it: Varies
- Coordinates: Varies

Messier 3 (25th) & Whirlpool Galaxy (12th)
as captured by the Hubble Space Telescope.

MOON

Ranked: 8th

Visibility rating: ★★★★★

Beauty	Power	Mystery	Wonder Rating
92	56	92	80

WHY IT IS IN THE LIST?

The Moon likely formed when a Mars-sized object hit the Earth. The debris from that impact eventually clumped together to become the Moon.

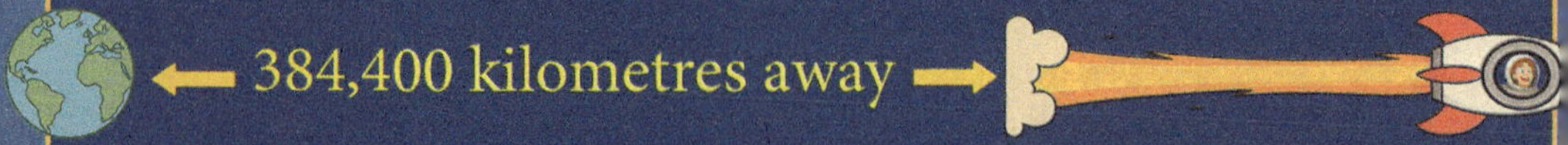

By far, one of the coolest cosmic coincidences is that our Moon is roughly 400x smaller than our Sun whilst also being 400x closer to the Earth than it. Leading to the perfect Solar eclipses that we have come to appreciate!

What You can See

The same Wonder captured through 3 VERY different setups

Seestar S50

$500

Remote observatory

$1,000,000

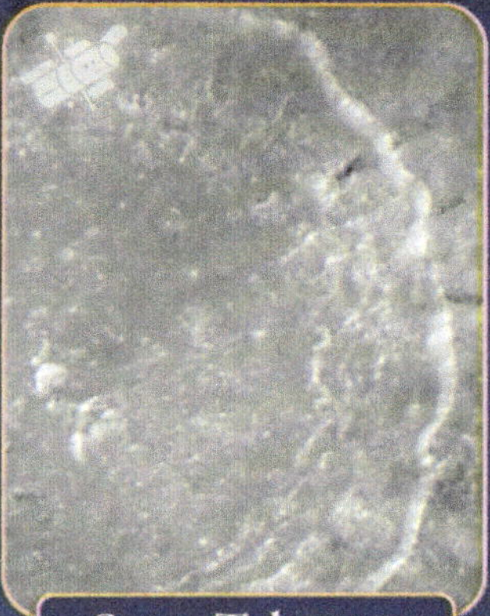

Space Telescope

$1,000,000,000

FUN FACT

The Moon is trying to get away from us, every year it is slowly drifting away from Earth at a rate of about 1.5 inches (3.8 cm). In our solar system Earth has the 6th highest number of Moons (1).

HOW TO SPOT IT

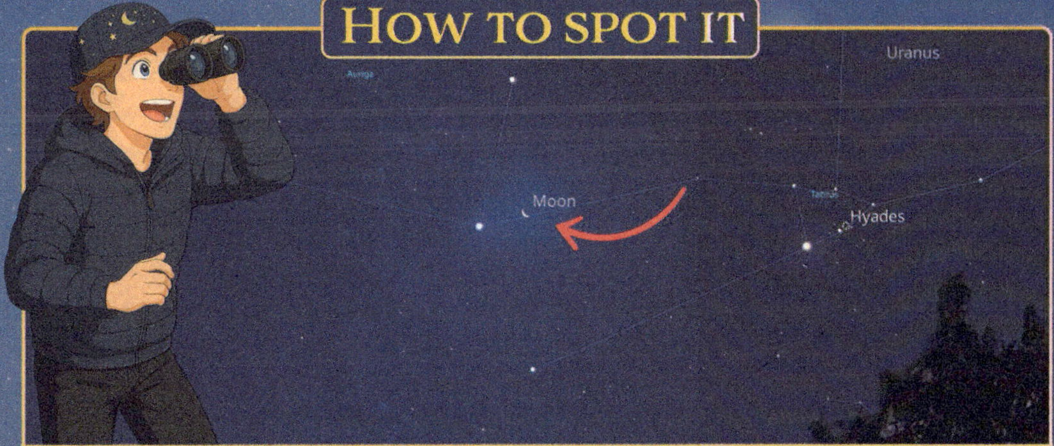

- Magnitude (Brightness): -12.7
- Visible to the naked eye: ☑
- Distance: 384,400km away
- Object catalog name: Moon/Luna

Have you spotted it yet? ☐

- Constellation: Varies
- Visibility rating: ★★★★★
- Best time to see it: Varies
- Coordinates: Varies

ALGOL

Ranked: 39th

Visibility rating: ★★★★★

Beauty	Power	Mystery	Wonder Rating
16	68	96	60

WHY IT IS IN THE LIST?

Algol, the "Demon Star", is famous for its regular dimming and brightening. This variability is caused by a larger, fainter companion star periodically passing in front of a brighter primary star, causing eclipses.

← 93 Light years away →

The brighter star in the system is actually younger, which is a bit weird. The reason behind this is that the older, less massive star has transferred mass to its companion. Basically, the younger star is what we call a vampire star, it steals matter from others...
Ancient astronomers referred to Algol and its dimming as a bad omen.

What You can See

The same Wonder captured through 3 VERY different setups

Seestar S50

$500

Remote observatory

$1,000,000

Space Telescope

$1,000,000,000

FUN FACT

Algol isn't just one star it's three stars orbiting each other. The two main stars eclipse each other, and the third orbits farther out.

HOW TO SPOT IT

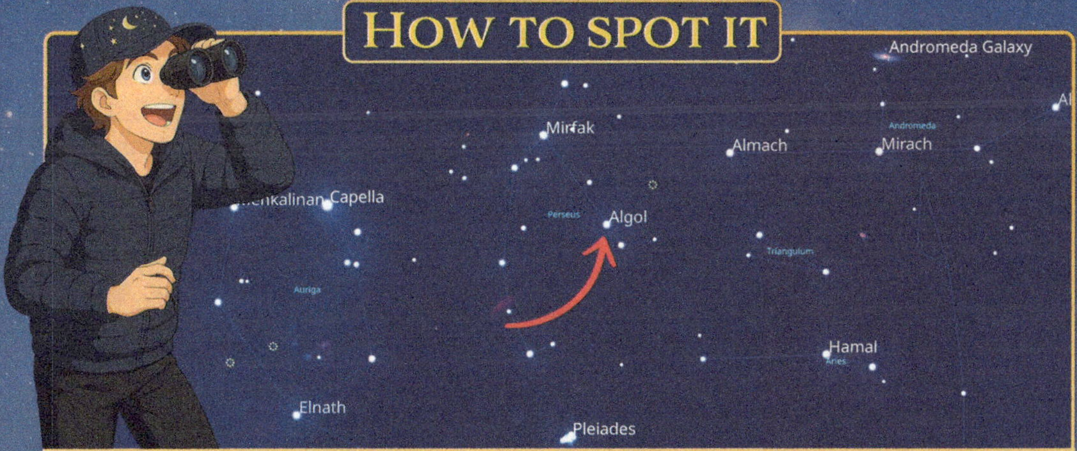

- Magnitude (Brightness): 2.1
- Visible to the naked eye: ☑
- Distance: 93 Light years
- Object catalog name: β Persei

Have you spotted it yet? ☐

- Constellation: Perseus
- Visibility rating: ★★★★★
- Best time to see it: December-Feb
- Coordinates: 03h 08m 10s, +40° 57′ 20″

Hind's Variable Nebula

Ranked: 23rd

Beauty	Power	Mystery	Wonder Rating
75	53	90	73

Why it is in the list?

The nebula is illuminated by a young star named T Tauri, which is throwing a massive temper tantrum causing the surrounding clouds of dust and gas to be randomly illuminated.

← 400 Light Years Away →

The official name for this deep sky object is "Hind's Variable Nebula" but I much prefer to refer to it as "The Devil in Disguise". Its phantom-like appearance causes your imagination to run wild with visions of cosmic monsters that feast on stars.

What You can See

The same Wonder captured through 3 VERY different setups

Remote observatory

Fun Fact

You are not seeing double. I really have used the same image of NGC 1555 twice. Why? Because it is HARD to image. Perhaps the most difficult target of the series, which is why neither Hubble or my $500 telescope have images to show for it!

How to spot it

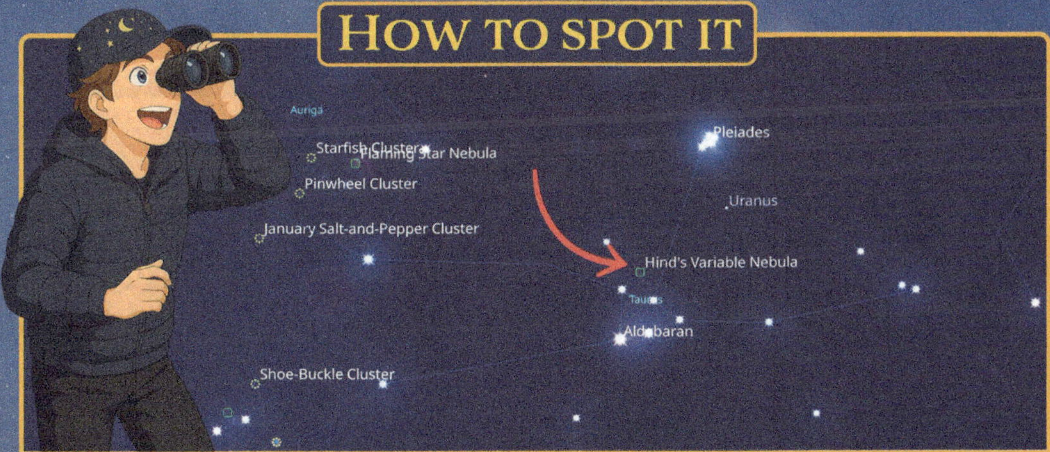

- Magnitude (Brightness): 9.98
- Visible to the naked eye: ☒
- Distance: 400 Light Years
- Object catalog name: NGC 1555

Have you spotted it yet? ☐

- Constellation: Taurus
- Visibility rating: ★☆☆☆☆
- Best time to see it: September-Feb
- Coordinates: 04h 21m 57s, +19° 32' 07"

FIREWORKS GALAXY

Ranked: 14th

Visibility rating: ★★☆☆☆

Beauty	Power	Mystery	Wonder Rating
53	98	83	78

WHY IT IS IN THE LIST?

NGC 6946 gets the name "Fireworks Galaxy" because of its excessively high rate of stellar birth and death. By observing this galaxy scientists learn how galaxies recycle material over time.

It sits right between the Milky Way and the IC 342 group of galaxies which means that it's partially obscured by dust in our own galaxy dimming it a bit, despite its relative brightness.

What You can See

The same Wonder captured through 3 VERY different setups

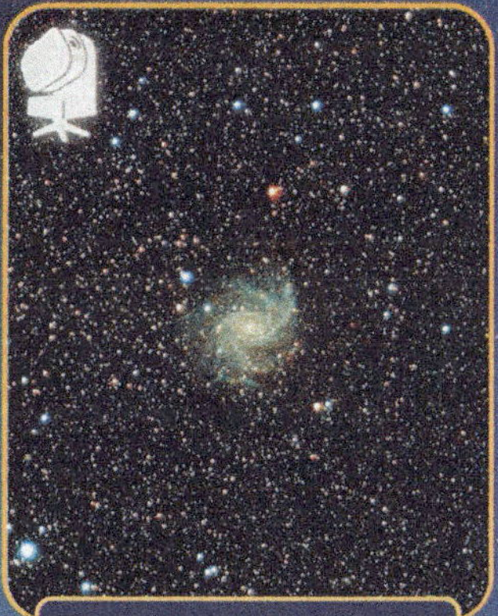

Seestar S50

$500

Remote observatory

$1,000,000

Space Telescope

$1,000,000,000

FUN FACT

On average we expect about 1 supernova to occur in a typical galaxy every 100 years. However over the last century or so there have been 10 different supernovae in the fireworks galaxy.

HOW TO SPOT IT

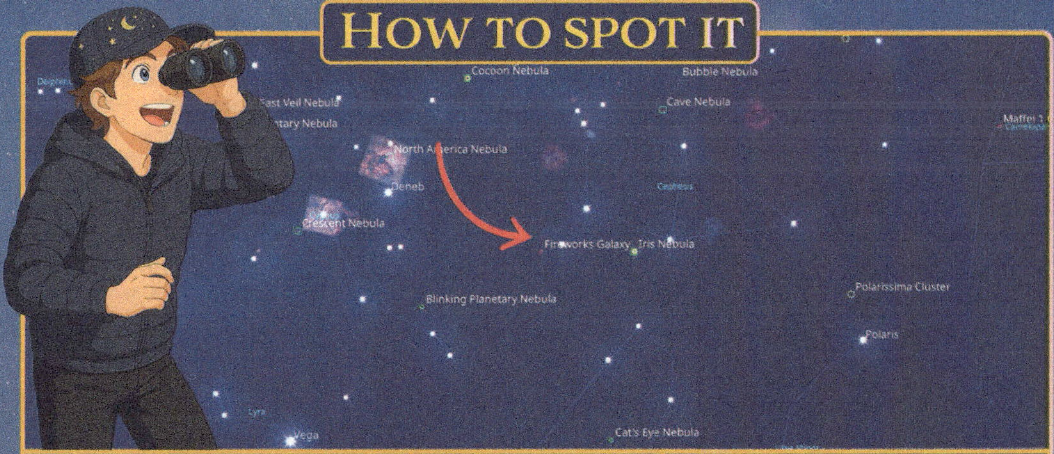

- Magnitude (Brightness): 9.6
- Visible to the naked eye: ☒
- Distance: 25,000,000 Light Years Away
- Object catalog name: NGC 6946

Have you spotted it yet? ☐

- Constellation: Cepheus
- Visibility rating: ★★☆☆☆
- Best time to see it: July-October
- Coordinates: 20h 34m 52s, +60° 09′ 14″

BETELGEUSE

Ranked: 19th

Visibility rating: ★★★★★

Beauty	Power	Mystery	Wonder Rating
52	87	85	75

WHY IT IS IN THE LIST?

It is usually the 10th brightest star, but its brightness changes unpredictably. That's because of its unstable nature causing it to expand and contract like a heart beat.

← 548 Light Years Away →

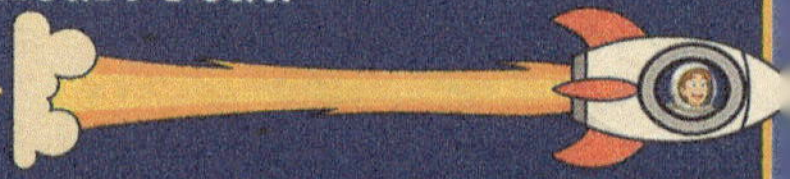

Betelgeuse is a red supergiant nearing the end of its life. Having already swollen to an enormous size, it is currently about 700 to 1,000 times the radius of our Sun. Betelgeuse is so large that if it replaced our Sun, its surface would extend beyond the orbits of Mercury, Venus, Earth, and Mars.

What You can See

The same Wonder captured through 3 VERY different setups

Seestar S50

$500

Remote observatory

$1,000,000

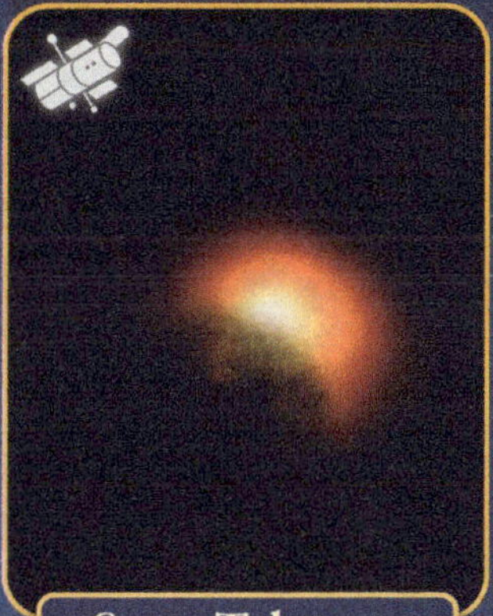

Space Telescope

$1,000,000,000

FUN FACT

By the time you are reading this Betelgeuse may have already gone supernova. When it does it could shine as bright as the full Moon for weeks. It will be so bright that it will likely remain in our daytime skies as a “second sun” for weeks.

HOW TO SPOT IT

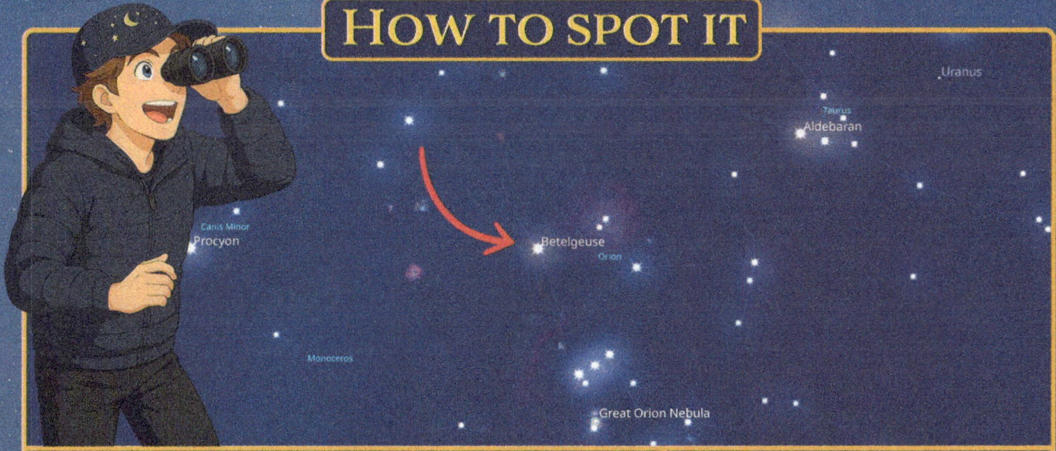

- Magnitude (Brightness): 0.42
- Visible to the naked eye: ☑
- Distance: 548 Light Years Away
- Object catalog name: Alpha Orionis

Have you spotted it yet? ☐

- Constellation: Orion
- Visibility rating: ★★★★★
- Best time to see it: December-Feb
- Coordinates: 05h 55m 10s, +07° 24′ 25″

BODE'S NEBULA & CIGAR GALAXY

Ranked: 33rd

Visibility rating: ★★★☆☆

Beauty	Power	Mystery	Wonder Rating
69	81	48	66

WHY IT IS IN THE LIST?

The Cigar galaxy (right) is known as a starburst galaxy because it is undergoing intense star formation about 10 times faster than that of our own Milky Way Galaxy.

←11,900,000 Light Years Away→

The Cigar galaxy is spewing massive superwinds of gas and dust out of its center. These winds are driven by supernovae and stellar winds from young, massive stars. Whilst Bode's nebula (left) is also in fact a galaxy despite the name first given to it, its edge on spiral appearance makes it an excellent sight through a telescope.

What You can See

The same Wonder captured through 3 VERY different setups

Seestar S50

$500

Remote observatory

$1,000,000

Space Telescope

$1,000,000,000

FUN FACT

Bode's nebula (M81) and the Cigar galaxy (M82) are locked in a gravitational encounter that has dramatically altered both galaxies. Their close pass ripped streams of gas between them, triggering explosive star formation.

HOW TO SPOT IT

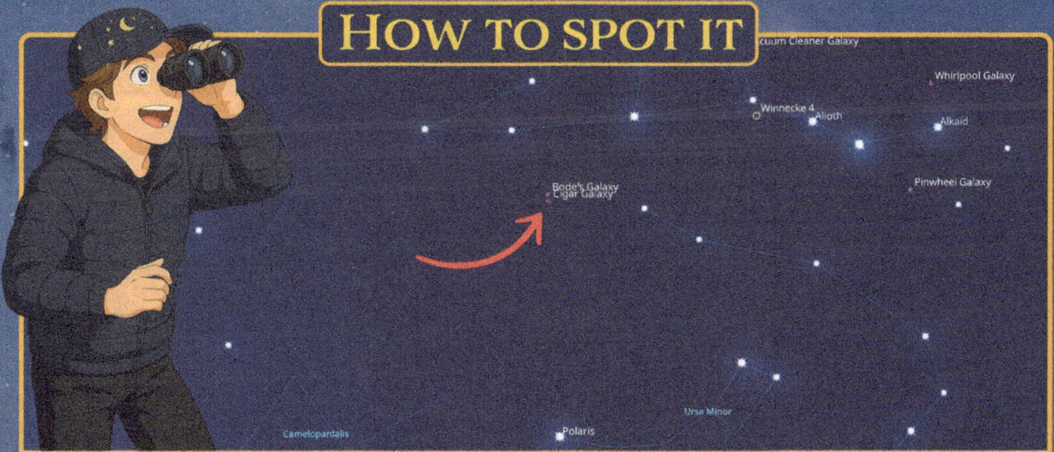

- Magnitude (Brightness): 8.4
- Visible to the naked eye: ☒
- Distance: 11,900,000 Light Years Away
- Object catalog name: M81/M82

Have you spotted it yet? ☐

- Constellation: Ursa Major
- Visibility rating: ★★★☆☆
- Best time to see it: March-May
- Coordinates: 09h 55m 52s, +69° 40′ 47″

PLEIADES

Ranked: 4th

Visibility rating: ★★★★★

Beauty	Power	Mystery	Wonder Rating
95	59	97	84

WHY IT IS IN THE LIST?

It is often mistaken for a small dipper because of its shape but it's not a constellation, in fact it is one of the brightest open star clusters visible in our night sky.

← 444 Light Years Away →

The nickname Seven Sisters is something of a mystery, to the naked eye it seems there are only 6 stars and yet many cultures referred to there being 7, creating stories and legends behind what happened to the 7th sister and why she is no longer visible to us in our night sky.

What You can See

The same Wonder captured through 3 VERY different setups

Seestar S50

$500

Remote observatory

$1,000,000

Space Telescope

$1,000,000,000

FUN FACT

While often called the Seven Sisters, it contains over a thousand visible stars, most of which are young, hot, and blue. The cluster is visible to the naked eye, with some observers able to see up to 14 stars under ideal conditions. How many can you spot?

HOW TO SPOT IT

- Magnitude (Brightness): 1.6
- Visible to the naked eye: ☑
- Distance: 444 Light Years Away
- Object catalog name: M45

Have you spotted it yet? ☐

- Constellation: Taurus
- Visibility rating: ★★★★★
- Best time to see it: November-March
- Coordinates: 03h 47m 24s, +24° 07' 00"

VENUS

Ranked: 38th

Visibility rating: ★★★★★

Beauty	Power	Mystery	Wonder Rating
66	37	79	61

WHY IT IS IN THE LIST?

Named after the goddess of love, Venus was once theorized to be home to alien life until a probe was sent to the surface of the planet and revealed a hellish and barren landscape.

⟵ 38,000,000 Kilometres Away ⟶

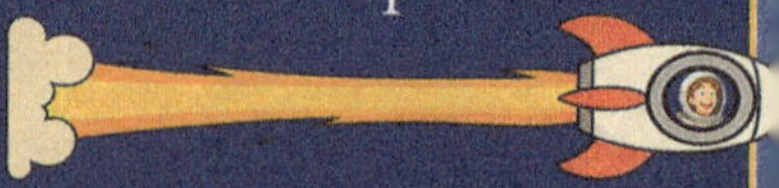

One day on Venus (one full spin) takes about 243 Earth days. But its year (one orbit around the Sun) is only 225 Earth days. So, a day is longer than a year on Venus!

Things get even weirder, Venus is the only planet in our solar system to rotate in the opposite direction.

What You can See

The same Wonder captured through 3 VERY different setups

Seestar S50

$500

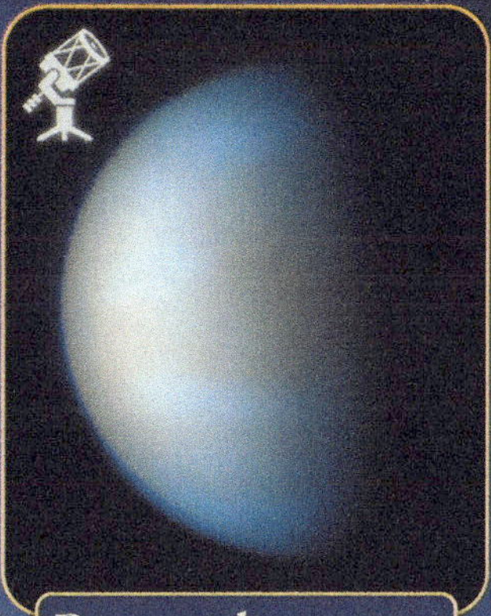

Remote observatory

$1,000,000

Space Telescope

$1,000,000,000

Fun Fact

It is the hottest planet in our solar system, despite not being the closest to the sun. This is because its dense carbon-dioxide atmosphere creates a runaway greenhouse effect that traps heat more efficiently than on any other planet.

How to spot it

- Magnitude (Brightness): -4.4
- Visible to the naked eye: ☑
- Distance: Closest 38,000,000 Km
- Object catalog name: Venus

Have you spotted it yet? ☐

- Constellation: Varies
- Visibility rating: ★★★★★
- Best time to see it: Varies
- Coordinates: Varies

Elephant Trunk Nebula

Ranked: 16th

Beauty	Power	Mystery	Wonder Rating
87	70	75	77

Why it is in the list?

Its most distinctive feature, the "trunk," is a dark, cold cloud of gas and dust, sculpted by the radiation and winds from nearby massive stars.

← 2,400 Light Years Away →

This stellar activity also triggers the birth of new stars within the nebula. Recent studies suggest that magnetic fields play a role in shaping the trunk-like structure. These invisible forces may help guide the flow of gas and affect how stars form.

What You can See

The same Wonder captured through 3 VERY different setups

Seestar S50

$500

Remote observatory

$1,000,000

Space Telescope

$1,000,000,000

Fun Fact

The Elephant Trunk nebula is a stellar nursery, a birthplace for new stars in our galaxy. The dense globule of dust that makes up the trunk hides our view of the stars within. But thanks to the help of infrared imaging with our space telescope we can see the births of new stars.

How to spot it

- Magnitude (Brightness): 5.6
- Visible to the naked eye: ☒
- Distance: 2,400 Light Years Away
- Object catalog name: IC 1396A

Have you spotted it yet? ☐

- Constellation: Cepheus
- Visibility rating: ★★★★☆
- Best time to see it: August-October
- Coordinates: 21h 35m 37s, +57° 24' 03"

The Pleiades (4th) & Elephant Trunk Nebula (16th) as captured by Telescope Live's remote observatory service.
Datasets start at less than $1.

NORTH AMERICA NEBULA

Ranked: 27th

Visibility rating: ★★★★☆

Beauty	Power	Mystery	Wonder Rating
88	60	65	71

WHY IT IS IN THE LIST?

The North America Nebula is one of the best deep sky objects in our night sky to image thanks to its bright colourful appearance. At its peak, the nebula appears overhead for northern observers.

← 2,590 Light Years Away →

The red glow of the nebula comes from ionized hydrogen, a common feature of emission nebulae.
The views achieved with our 3 different setups showcase the nebula in visible colour, narrowband and infrared.
Providing 3 unique perspectives of this enormous nebula.

What You can See

The same Wonder captured through 3 VERY different setups

Seestar S50

$500

Remote observatory

$1,000,000

Space Telescope

$1,000,000,000

Fun Fact

With a diameter of 100 Light Years, this nebula occupies a diameter in our night sky four times larger than the full Moon. Which is huge but its nebulosity is still too faint to be seen fully with the naked eye.

How to spot it

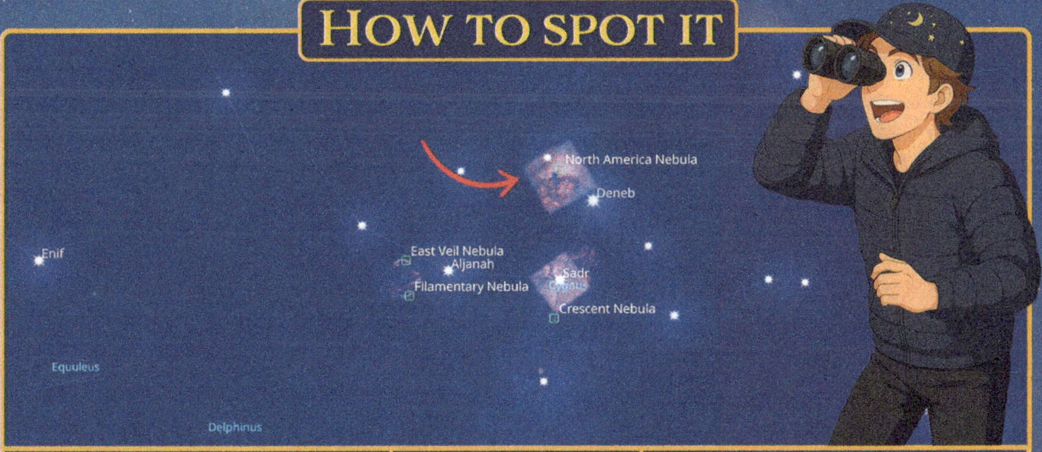

- Magnitude (Brightness): 4
- Visible to the naked eye: ☒
- Distance: 2,590 Light Years Away
- Object catalog name: NGC 7000

Have you spotted it yet? ☐

- Constellation: Cygnus
- Visibility rating: ★★★★☆
- Best time to see it: June-October
- Coordinates: 20h 58m 54s, +44° 19′ 00″

MIZAR & ALCOR

Ranked: 20th

Visibility rating: ★★★★★

Beauty	Power	Mystery	Wonder Rating
64	77	82	74

WHY IT IS IN THE LIST?

What looks like a simple pair of stars is actually a complex multi-star system, revealing how deceptive and layered the night sky really is.

← 83 & 81 Light Years Away →

For keen-eyed observers, you can spot two stars making up the second star in from the handle of Ursa Major. But the truth is, Mizar and Alcor are only a small part of a sextuplet star system. That's six stars all orbiting one another. Can you imagine how chaotic life would be there? Six sunsets a day doesn't sound too bad, though!

What You can See

The same Wonder captured through 3 VERY different setups

Seestar S50

$500

Remote observatory

$1,000,000

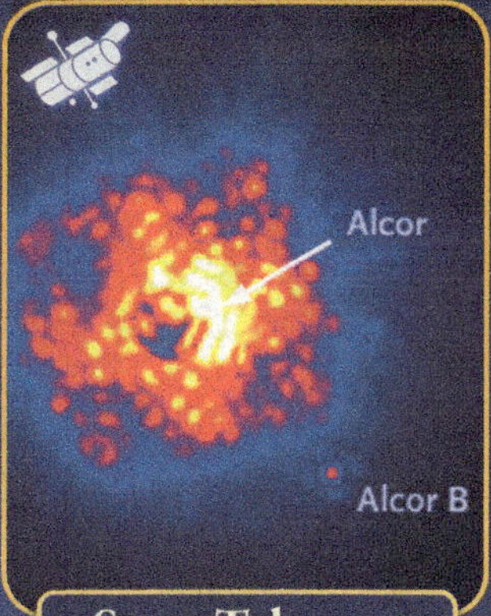

Space Telescope

$1,000,000,000

FUN FACT

They are gravitationally bound and part of the Ursa Major Moving Group. Historically, they were used as a vision test, with the ability to distinguish them indicating that you had good eyesight. Can you spot them?

HOW TO SPOT IT

- Magnitude (Brightness): 2.2 & 4
- Visible to the naked eye: ☑
- Distance: 83 Light Years Away
- Object catalog name: Mizar & Alcor

Have you spotted it yet? ☐

- Constellation: Ursa Major
- Visibility rating: ★★★★★
- Best time to see it: March-June
- Coordinates: 13h 23m 55s, +54° 55 31"

Whale Galaxy

Ranked: 35th

Visibility rating: ★☆☆☆☆

Beauty	Power	Mystery	Wonder Rating
70	78	48	65

Why it is in the list?

When ticking off your galactic bucket list, NGC 4631 is an absolute must, its peculiar shape makes it a very unique sight in our night sky.

←25,000,000 Light Years Away→

Spring is the ideal season to go galaxy hunting, nearby to the Whale Galaxy you can also find the Sunflower, Crowbar, Black eye and Crocs eye Galaxies. Each with their own unique shapes.

What You can See

The same Wonder captured through 3 VERY different setups

Seestar S50

$500

Remote observatory

$1,000,000

Space Telescope

$1,000,000,000

Fun Fact

The Whale galaxy isn't drifting through space all alone, it has a smaller companion galaxy, NGC 4627, sometimes called the "Pup", which is interacting with it gravitationally and therefore partnering it in their journey through the cosmos.

How to spot it

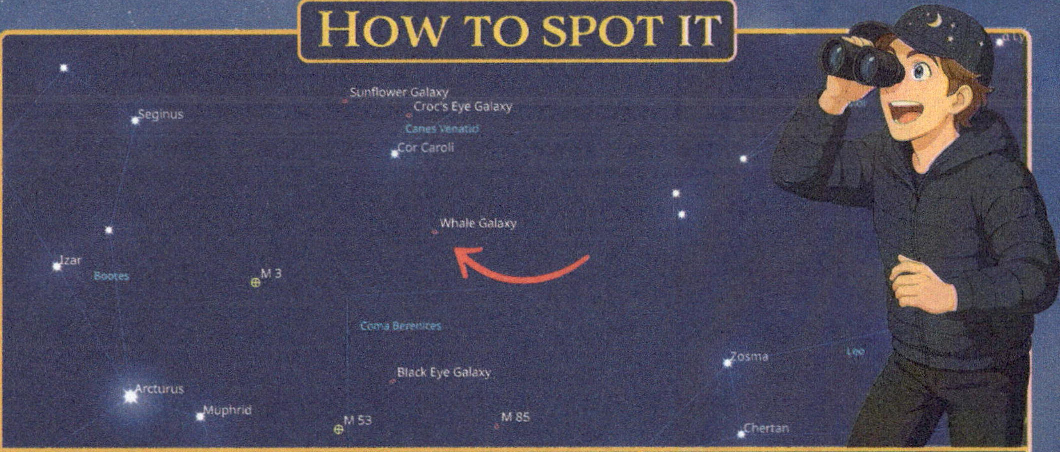

- Magnitude (Brightness): 9.8
- Visible to the naked eye: ☒
- Distance: 25,000,000 Light Years Away
- Object catalog name: NGC 4631

Have you spotted it yet? ☐

- Constellation: Canes Venatici
- Visibility rating: ★☆☆☆☆
- Best time to see it: March-May
- Coordinates: 12h 42m 08s, +32° 32' 30"

Pillars of Creation

Ranked: 7th

Visibility rating: ★★★☆☆

Beauty	Power	Mystery	Wonder Rating
93	71	79	81

Why it is in the list?

This may be the most iconic deep sky object in history. The Pillars of Creation are the perfect embodiment of everything that makes stargazing so exciting.

← 7,000 Light Years Away →

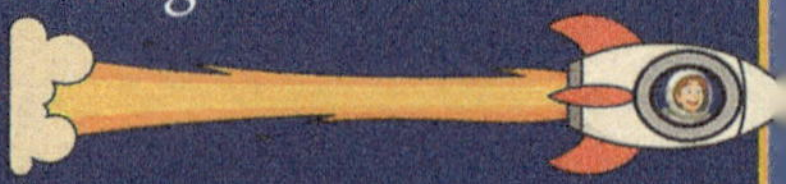

These towering stellar nurseries are an epic reminder of how small we are on the grand scale of the universe. The Hubble image from 1995 became one of the most iconic space photos ever taken. It revealed immense structures of dust and gas silhouetted against a glowing background.

What You can See

The same Wonder captured through 3 VERY different setups

Seestar S50

$500

Remote observatory

$1,000,000

Space Telescope

$1,000,000,000

FUN FACT

They're being destroyed as we speak. The same radiation that makes them glow is also eroding them, so just like everything else in our universe, the Pillars of Creation are also temporary.

HOW TO SPOT IT

- Magnitude (Brightness): 6.0
- Visible to the naked eye: ☒
- Distance: 7,000 Light Years Away
- Object catalog name: M16/NGC 6611

Have you spotted it yet? ☐

- Constellation: Serpens
- Visibility rating: ★★★ (3 of 5 stars)
- Best time to see it: June-August
- Coordinates: 18h 18m 48s, -13° 49 00″

NEEDLE GALAXY

Ranked: 17th

Visibility rating: ★★☆☆☆

Beauty	Power	Mystery	Wonder Rating
93	71	79	81

WHY IT IS IN THE LIST?

It's tilted at just about 3 degrees from exactly edge-on, making it one of the flattest-looking galaxies in the sky. That rare angle gives it its iconic, blade-like appearance.

←40,000,000 Light Years Away→

The shape and features suggest it may have absorbed smaller galaxies in the past. These interactions often leave subtle traces in the halo or star distribution.
Could the Needle Galaxy be a serial galactic cannibal?

What You can See

The same Wonder captured through 3 VERY different setups

Seestar S50

$500

Remote observatory

$1,000,000

Space Telescope

$1,000,000,000

FUN FACT

The bluish colour that can be seen along the disc gives evidence of the spiral nature of the galaxy, indicating the presence of hot, luminous, young stars, born out of clouds of interstellar gas.

HOW TO SPOT IT

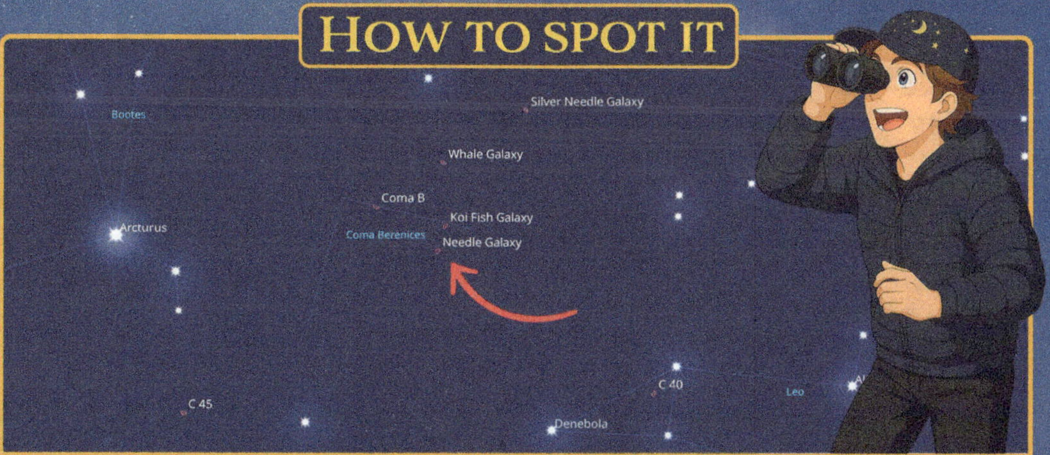

- Magnitude (Brightness): 10.4
- Visible to the naked eye: ☒
- Distance: 40,000,000 Light Years Away
- Object catalog name: NGC 4565

Have you spotted it yet? ☐

- Constellation: Coma Berenices
- Visibility rating: ★★☆☆☆
- Best time to see it: March-May
- Coordinates: 12h 36m 21s, -25° 59 15"

Rosette Nebula

Ranked: 28th

Beauty	Power	Mystery	Wonder Rating
82	70	60	71

Why it is in the list?

The Rosette Nebula gets its name because it resembles a blooming flower, a rosette especially in deep-sky photographs.

← 5,200 Light Years Away →
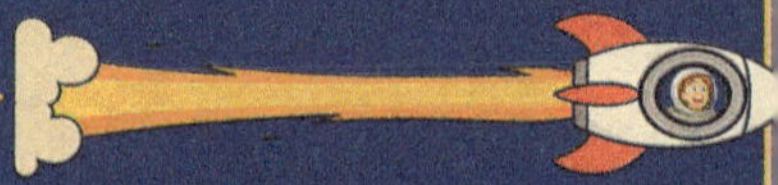

Most of its light comes from ionized hydrogen gas, which gives it that deep red or pinkish glow often seen in astrophotography. If it were as close as the Orion Nebula (1,350 light years) it would take up the same amount of space in our night sky as 10 full moons, it is huge!

What You can See

The same Wonder captured through 3 VERY different setups

Seestar S50

$500

Remote observatory

$1,000,000

Space Telescope

$1,000,000,000

Fun Fact

The young stars in the nebula's core emit powerful stellar winds that carve out the cavity and shape the surrounding gas and dust into petal-like structures, it looks gentle to the eye but really this is a very violent landscape.

How to spot it

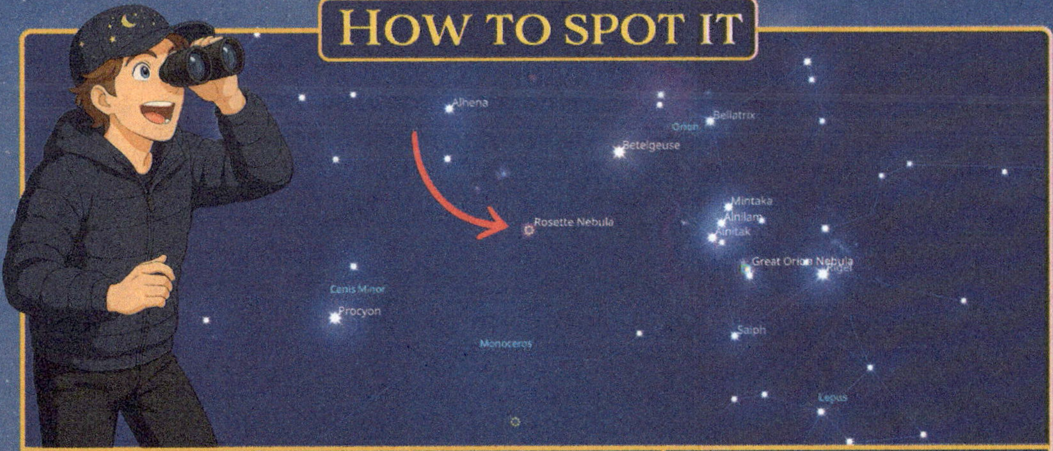

- Magnitude (Brightness): 9
- Visible to the naked eye: ☒
- Distance: 5,200 Light Years Away
- Object catalog name: NGC 2244

Have you spotted it yet? ☐

- Constellation: Monoceros
- Visibility rating: ★★★☆☆
- Best time to see it: December-February
- Coordinates: 06h 33m 45s, +04° 59 54"

Veil Nebula

Ranked: 15th

Beauty	Power	Mystery	Wonder Rating
87	72	74	78

Why it is in the list?

The veil nebula is the remnant of a supernova explosion, shockwaves from the original explosion are still moving outward at about 1.5 million kilometers per hour through space.

← 2,400 Light Years Away →

Though to us, the nebula seems "frozen" in space because of its vast size and distance.
Astrophotographers categorise the nebula into multiple individual deep sky objects, including the Eastern and Western Veil Nebula.

What You can See

The same Wonder captured through 3 VERY different setups

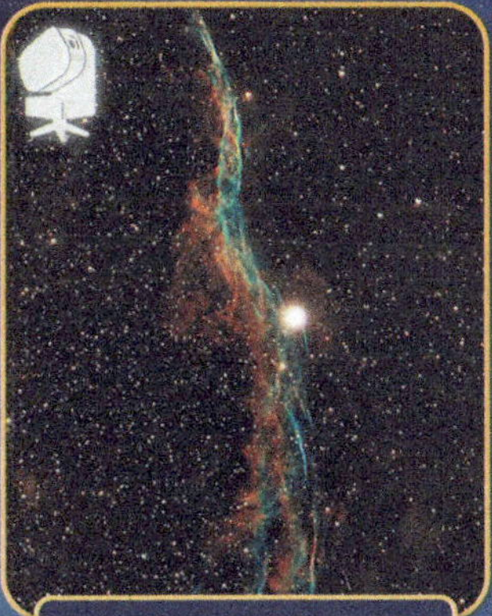

Seestar S50

$500

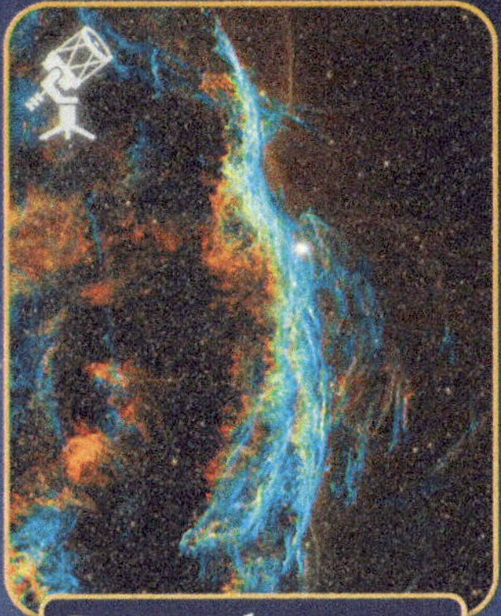

Remote observatory

$1,000,000

Space Telescope

$1,000,000,000

Fun Fact

The nebula covers the same area in our night sky as 6 full moons, it is huge! It's estimated to have gone supernova 10,000 years ago which means there's a good chance early humans observed its explosive death.

How to spot it

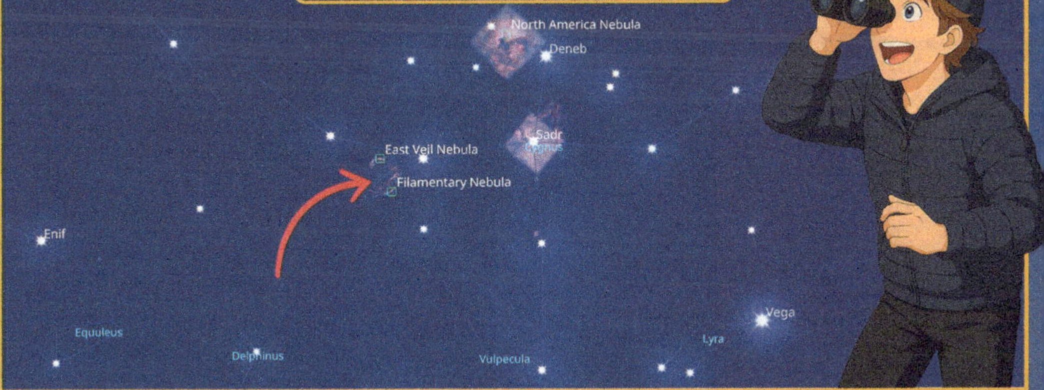

- Magnitude (Brightness): 7
- Visible to the naked eye: ☒
- Distance: 2,400 Light Years Away
- Object catalog name: NGC 6960

Have you spotted it yet? ☐

- Constellation: Cygnus
- Visibility rating: ★★★☆☆
- Best time to see it: July-October
- Coordinates: 20h 45m 38s, +30° 42 30"

Pinwheel Galaxy

Ranked: 32nd

Visibility rating: ★★☆☆☆

Beauty	Power	Mystery	Wonder Rating
70	81	50	67

Why it is in the list?

It is considered a flocculent spiral galaxy, meaning its spiral arms are patchy, fluffy, and less well-defined compared to galaxies like the Whirlpool Galaxy.

←21,000,000 Light Years Away→

The visible mass of M101 can't account for its rotation speed, this suggests it's surrounded by a massive halo of dark matter, helping hold it together gravitationally.

What You can See

The same Wonder captured through 3 VERY different setups

Seestar S50

$500

Remote observatory

$1,000,000

Space Telescope

$1,000,000,000

Fun Fact

As recently as 2023, we have observed supernova explosions within the Pinwheel Galaxy. At a magnitude off 11, even amateur astronomers could see the explosion from their back gardens with their own amateur setups.

How to spot it

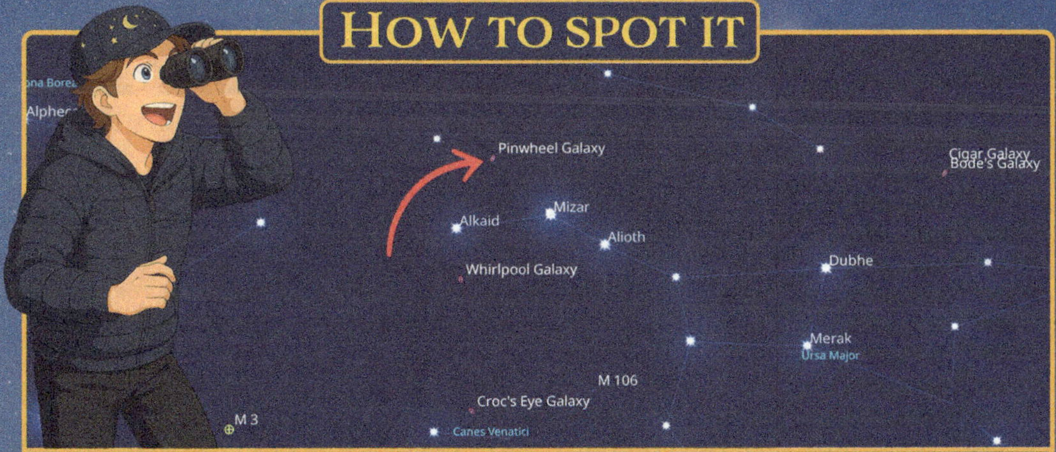

- Magnitude (Brightness): 7.9
- Visible to the naked eye: ☒
- Distance: 21,000,000 Light Years Away
- Object catalog name: M101/NGC 5457

Have you spotted it yet? ☐

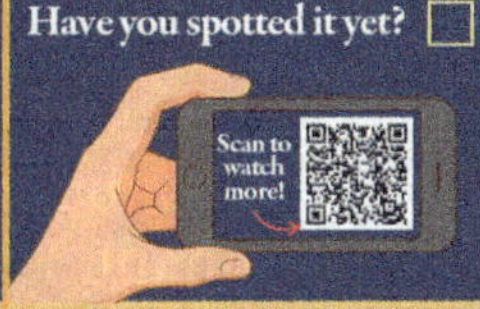

- Constellation: Ursa Major
- Visibility rating: ★★☆☆☆
- Best time to see it: March-May
- Coordinates: 14h 03m 13s, +54 20 57"

Crab Nebula

Ranked: 1st

Beauty	Power	Mystery	Wonder Rating
65	97	98	87

Why it is in the list?

The name "Crab" came from 19th-century astronomers who thought its shape resembled a crab in early sketches.
Now that we have detailed long exposure images you can decide for yourself what you think it really looks like... An egg?

6,500 Light Years Away

Even though it has been almost a 1,000 years since the crab nebula first appeared in our night sky, we can still see it evolving every year. In fact, you can even compare your shots with professional images captured decades ago to show the dramatic expansion of this explosive remnant.

What You can See

The same Wonder captured through 3 VERY different setups

Seestar S50

$500

Remote observatory

$1,000,000

Space Telescope

$1,000,000,000

Fun Fact

When the star at the centre of the Crab Nebula went supernova in the year 1054AD, there were multiple documented observations of a bright point of light appearing in our night sky that remained visible during the daytime for 23 days before it started to fade.

How to spot it

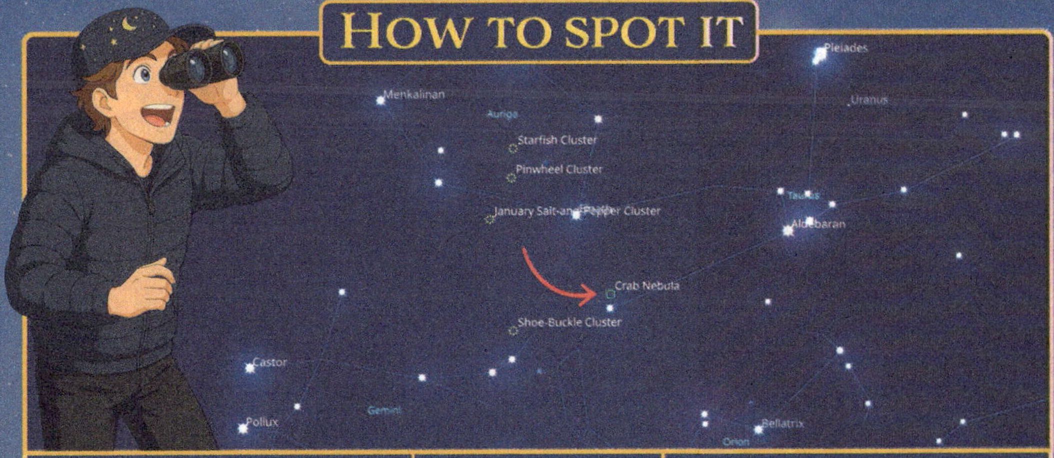

- Magnitude (Brightness): 8.4
- Visible to the naked eye: ☒
- Distance: 6,500 Light Years Away
- Object catalog name: M1/NGC 1952

Have you spotted it yet? ☐

- Constellation: Taurus
- Visibility rating: ★★☆☆☆
- Best time to see it: November-Feb
- Coordinates: 05h 34m 32s, +22° 00 52"

THE WONDERWALL

42 WONDERS OF OUR NIGHT SKY

#	Wonder	Score
1.	Crab Nebula	87
2.	Andromeda Galaxy	86
3.	Orion Nebula	85
4.	Pleiades	84
5.	Saturn	83
6.	Jupiter	81
7.	Pillars of Creation	81
8.	Moon	80
9.	Albireo	80
10.	Sun	79
11.	Sombrero Galaxy	79
12.	Whirlpool Galaxy	79
13.	Helix Nebula	78
14.	Fireworks Galaxy	78
15.	Veil Nebula	78
16.	Elephant Trunk	77
17.	Needle Galaxy	77
18.	Antennae Galaxy	76
19.	Betelgeuse	75
20.	Mizar & Alcor	74
21.	Horsehead Nebula	74
22.	Hubble's V-Nebula	73
23.	Hinds V-Nebula	73
24.	Mars	73
25.	M87	72
26.	Messier 3	72
27.	North America	71
28.	Rosette Nebula	71
29.	Hercules Cluster	70
30.	Leo Triplet	69
31.	Heart Nebula	68
32.	Pinwheel Galaxy	67
33.	M81 & M82	66
34.	Ring Nebula	66
35.	Whale Galaxy	65
36.	Thor's Helmet	63
37.	Dumbbell Nebula	62
38.	Venus	61
39.	Algol	60
40.	Beehive Cluster	51
41.	ISS	50
42.	Uranus	49

The Answer to the Ultimate Question of Life, the Universe, and Everything...

Top 10 Wonders Ranked by Beauty

1.	Saturn		97
1.	Antennae Galaxy		97
3.	Albireo		96
3.	Helix Nebula	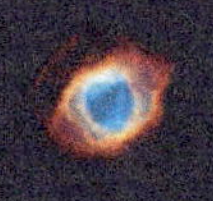	96
3.	Heart Nebula		96
6.	Orion Nebula		95
6.	Pleiades		95
8.	Pillars of Creation		93
9.	Moon		92
10.	Whirlpool Galaxy		92

Top 10 Wonders Ranked by Power

1.	M87		100
2.	Fireworks Galaxy		98
3.	Crab Nebula		97
4.	Whirlpool Galaxy		94
5.	Sun	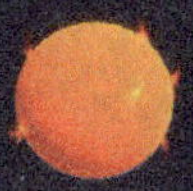	91
6.	Antennae Galaxy		88
7.	Betelgeuse	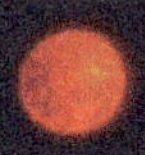	87
8.	Leo Triplet		86
9.	M81 & M82		81
10.	Pinwheel Galaxy		81

Top 10 Wonders Ranked by Mystery

1.	M87		100
2.	Andromeda Galaxy		99
3.	Crab Nebula		98
4.	Pleiades		97
5.	Algol	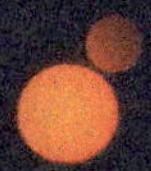	96
6.	Moon		92
7.	Hinds V-Nebula		90
8.	Orion Nebula		89
9.	Hubble's V-Nebula		86
10.	Betelgeuse	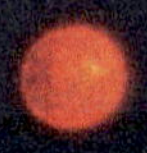	85

Top 10 Wonders Ranked by YOU

1.
2.
3.
4.
5.
6.
7.
8.
9.
10.

DOLPHIN NEBULA

BONUS

Visibility rating: ★★☆☆☆

Beauty	Power	Mystery	Wonder Rating
99	56	59	78

HOW TO SPOT IT

- Magnitude (Brightness): 7.0
- Visible to the naked eye: ☒
- Distance: 4,500 Light Years
- Object catalog name: Sh2-308

Have you spotted it yet? ☐

- Constellation: Canis Major
- Visibility rating: ★★☆☆☆
- Best time to see it: December - March
- Coordinates: 06h 54m 13s, -23° 55′ 42″

SIRIUS

BONUS

Visibility rating: ★★★★★

Beauty	Power	Mystery	Wonder Rating
82	55	61	66

HOW TO SPOT IT

- Magnitude (Brightness): -1.46
- Visible to the naked eye: ☑
- Distance: 8.6 Light Years
- Object catalog name: α Canis Majoris

Have you spotted it yet? ☐

Scan to watch more!

- Constellation: Canis Major
- Visibility rating: ★★★★★
- Best time to see it: December - March
- Coordinates: 06h 45m 11s, -16° 42′ 05″

TRIANGULUM GALAXY

BONUS

Visibility rating: ★★★☆☆

Beauty	Power	Mystery	Wonder Rating
57	75	49	60

HOW TO SPOT IT

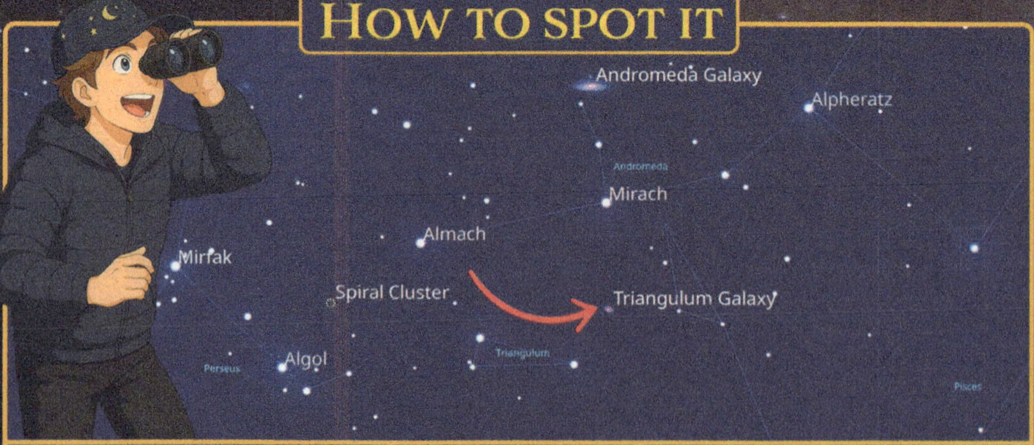

- Magnitude (Brightness): 5.7
- Visible to the naked eye: ☑
- Distance: 2,880,000 Light Years
- Object catalog name: M33 / NGC 598

Have you spotted it yet? ☐

- Constellation: Triangulum
- Visibility rating: ★★★☆☆
- Best time to see it: August - December
- Coordinates: 01h 33m 50s, +30° 39′ 36″

BLUE HORSEHEAD NEBULA

BONUS

Visibility rating: ★★★★☆

Beauty	Power	Mystery	Wonder Rating
91	56	59	69

HOW TO SPOT IT

- Magnitude (Brightness): 3.9
- Visible to the naked eye: ☑
- Distance: 424 Light Years
- Object catalog name: IC 4592

Have you spotted it yet? ☐

- Constellation: Scorpius
- Visibility rating: ★★★★☆
- Best time to see it: May - August
- Coordinates: 16h 12m 59s, -19° 18' 00"

Monkey Head Nebula

BONUS

Visibility rating: ★★★☆☆

Beauty	Power	Mystery	Wonder Rating
76	63	51	63

How to spot it

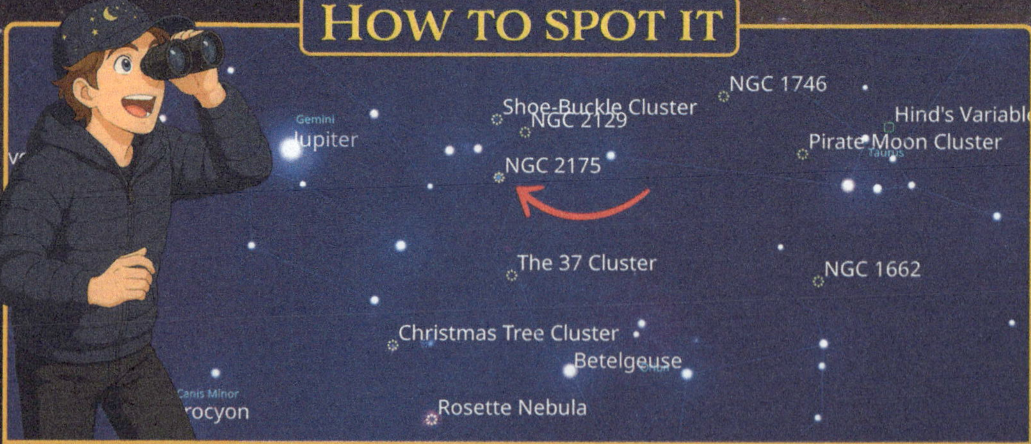

- Magnitude (Brightness): 6.8
- Visible to the naked eye: ☒
- Distance: 6,400 Light Years
- Object catalog name: NGC 2174

Have you spotted it yet? ☐

Scan to watch more!

- Constellation: Orion
- Visibility rating: ★★★☆☆
- Best time to see it: December - March
- Coordinates: 06h 09m 41s, +20° 29' 19"

CHRISTMAS TREE CLUSTER

BONUS

Visibility rating: ★★★★☆ (4 of 5 stars)

Beauty	Power	Mystery	Wonder Rating
82	55	61	66

HOW TO SPOT IT

- Magnitude (Brightness): 3.9
- Visible to the naked eye: ☑
- Distance: 2,400 Light Years
- Object catalog name: NGC 2264

Have you spotted it yet? ☐

- Constellation: Monoceros
- Visibility rating: ★★★★☆ (4 of 5 stars)
- Best time to see it: December - March
- Coordinates: 06h 41m 06s, +09° 53' 01"

Croc's Eye Galaxy

BONUS

Visibility rating: ★★☆☆☆

Beauty	Power	Mystery	Wonder Rating
61	77	46	61

How to spot it

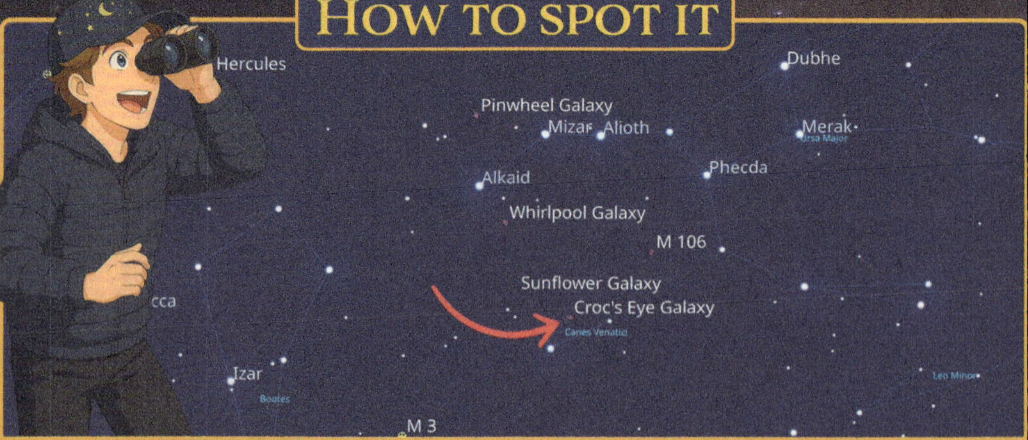

- Magnitude (Brightness): 8.2
- Visible to the naked eye: ☒
- Distance: 16,000,000 Light Years
- Object catalog name: M94/NGC 4736

Have you spotted it yet? ☐

- Constellation: Canes Venatici
- Visibility rating: ★★☆☆☆
- Best time to see it: April - June
- Coordinates: 12h 50m 53s, +41° 07' 12"

LAGOON NEBULA

BONUS

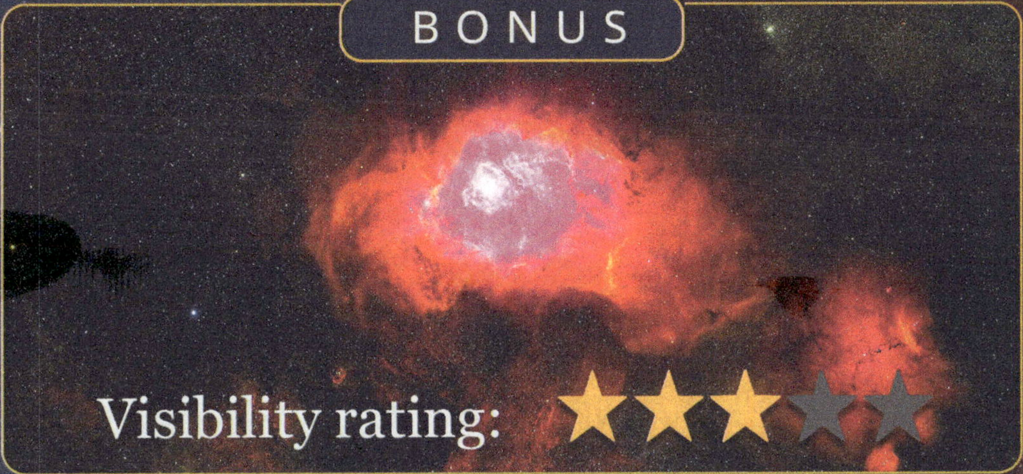

Visibility rating: ★★★☆☆

Beauty	Power	Mystery	Wonder Rating
86	71	67	75

HOW TO SPOT IT

- Magnitude (Brightness): 6.0
- Visible to the naked eye: ☒
- Distance: 5,200 Light Years
- Object catalog name: M8 / NGC 6523

Have you spotted it yet? ☐

- Constellation: Sagittarius
- Visibility rating: ★★★☆☆
- Best time to see it: June - August
- Coordinates: 18h 03m 41s, -24° 22' 49"

SEAGULL NEBULA

BONUS

Visibility rating: ★☆☆☆☆

Beauty	Power	Mystery	Wonder Rating
80	56	54	63

HOW TO SPOT IT

- Magnitude (Brightness): 15.2
- Visible to the naked eye: ☒
- Distance: 3,650 Light Years
- Object catalog name: NGC 2177

Have you spotted it yet? ☐

- Constellation: Monoceros
- Visibility rating: ★☆☆☆☆
- Best time to see it: December - March
- Coordinates: 07h 04m 25s, -10° 27' 18"

MARKARIAN CHAIN

BONUS

Visibility rating: ★★☆☆☆

Beauty	Power	Mystery	Wonder Rating
69	96	61	75

HOW TO SPOT IT

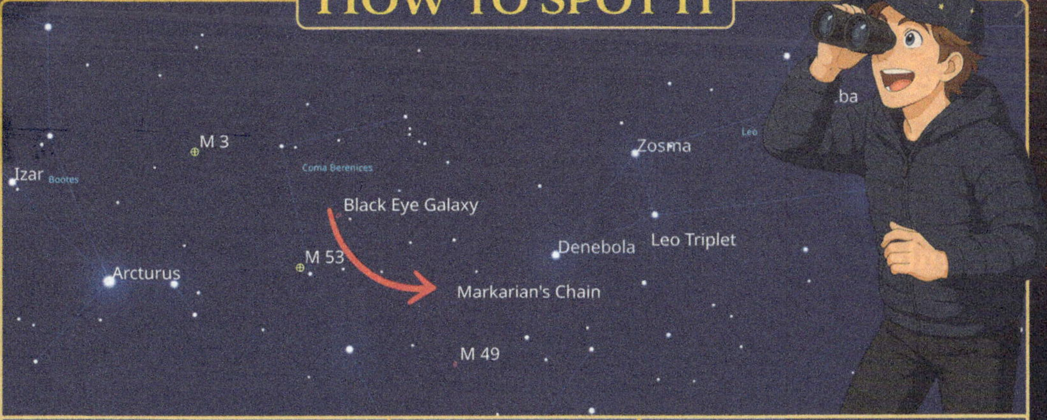

- Magnitude (Brightness): 9.0
- Visible to the naked eye: ☒
- Distance: 55,000,000 Light Years
- Object catalog name: M84/M86

Have you spotted it yet? ☐

Scan to watch more!

- Constellation: Virgo
- Visibility rating: ★★☆☆☆
- Best time to see it: March - June
- Coordinates: 12h 25m 03s, +12° 53′ 13″

42 Wonders Checklist

How many of the wonders have you spotted so far?

- ☐ Albireo
- ☐ Algol
- ☐ Andromeda Galaxy
- ☐ Antennae Galaxy
- ☐ Beehive Cluster
- ☐ Betelgeuse
- ☐ Crab Nebula
- ☐ Cigar Galaxy
- ☐ Dumbbell Nebula
- ☐ Elephant Trunk Neb
- ☐ Fireworks Galaxy
- ☐ Heart Nebula
- ☐ Helix Nebula
- ☐ Hercules Cluster
- ☐ Hind's V-Nebula
- ☐ Horsehead Nebula
- ☐ Hubble's V-Nebula
- ☐ ISS
- ☐ Jupiter
- ☐ Leo Triplet
- ☐ M87 (Black Hole)
- ☐ Mars
- ☐ Messier 3
- ☐ Mizar & Alcor
- ☐ Moon
- ☐ Needle Galaxy
- ☐ North America Neb
- ☐ Orion Nebula
- ☐ Pillars of Creation
- ☐ Pinwheel Galaxy
- ☐ Pleiades
- ☐ Ring Nebula
- ☐ Rosette Nebula
- ☐ Saturn
- ☐ Sombrero Galaxy
- ☐ Sun
- ☐ Thor's Helmet
- ☐ Uranus
- ☐ Veil Nebula
- ☐ Venus
- ☐ Whale Galaxy
- ☐ Whirlpool Galaxy

Sketch what you see

Draw what each of these deep sky objects looks like through your telescope.

Pleiades (M45)

Beehive Cluster (M44)

Orion Nebula (M42)

Andromeda Galaxy (M31)

Rosette Nebula (C49)

Hercules Cluster (M13)

When to spot them?

Spring (March – May)

Antennae Galaxy
Beehive Cluster
Croc's Eye Galaxy
Cigar Galaxy
Leo Triplet
M87 (Black Hole)
Markarian Chain
Messier 3
Mizar & Alcor
Needle Galaxy
Pinwheel Galaxy
Sombrero Galaxy
Whale Galaxy
Whirlpool Galaxy

Summer (June – August)

Albireo
Blue Horsehead Nebula
Dumbbell Nebula
Hercules Globular Cluster
Lagoon Nebula
North America Nebula
Pillars of Creation
Ring Nebula
Veil Nebula

Autumn (September – November)

Algol
Andromeda Galaxy
Elephant Trunk Nebula
Fireworks Galaxy
Heart Nebula
Helix Nebula
Hind's Variable Nebula
Triangulum Galaxy

Winter (December – February)

Betelgeuse
Christmas Tree Cluster
Crab Nebula
Horsehead Nebula
Hubble's Variable Nebula
Orion Nebula
Pleiades
Rosette Nebula
Sirius
Thor's Helmet

Varies (Year Round)

International Space Station
Jupiter, Mars, Moon, Saturn, Sun, Uranus, Venus

Astronomical darkness is the best window for stargazing and imaging. It's when the Sun is more than 18° below the horizon and the sky is properly dark. This table shows the average hours of astronomical darkness each month (rounded to the nearest 0.5 hour) for London, New York, Miami, and Sydney, highlighting how dramatically it changes with season and latitude. **(Note how during June & July London gets 0 hours!)**

City \ Month	Jan	Feb	Mar	Apr	May	Jun	Jul	Aug	Sep	Oct	Nov	Dec
London	11.5h	10.5h	8.5h	6.0h	2.5h	0.0h	0.0h	4.5h	7.5h	9.5h	11.0h	12.0h
New York	11.0h	10.5h	9.0h	7.5h	6.0h	4.5h	5.0h	6.5h	8.5h	10.0h	11.0h	11.5h
Miami	10.5h	10.0h	9.5h	8.5h	8.0h	7.5h	7.5h	8.0h	9.0h	10.0h	10.5h	10.5h
Sydney	6.5h	7.5h	9.0h	10.0h	10.5h	11.0h	11.0h	10.5h	9.5h	8.0h	7.0h	6.0h

42 WONDERS OF OUR NIGHT SKY
BINGO

SPOTTED THE SPACE STATION (ISS)	NECK ACHE FROM LOOKING UP FOR SO LONG	SPOTTED A SHOOTING STAR	CLOUDS STOPPED YOU FROM STARGAZING	SPOTTED A PLANET
FORGOT YOU LEFT THE LENS CAP ON TELESCOPE /CAMERA	DOG BARKED AT NOTHING	KICKED TRIPOD DURING AN IMAGE EXPOSURE	FOUND ALL 10 DIFFERENCES IN THE SPOT PUZZLE	SPOTTED THE SEVEN SISTERS
SCANNED A QR CODE TO WATCH A VIDEO EPISODE	SPOTTED THE DEMON STAR	FREE SPACE	FORGOT TO TURN BRIGHTNESS OF PHONE DOWN. NOW BLIND.	MADE YOURSELF A WARM CUP OF TEA / COFFEE
SPOTTED ALBIREO THROUGH A TELESCOPE	HEARD SOME SPOOKY NOISES	FOUND THE BIG DIPPER (URSA MAJOR)	CHECKED TO SEE IF IT WILL RAIN TONIGHT	SOLVED THE MEANING OF LIFE PUZZLE
DISAGREED WITH ONE OF MY WONDER RATINGS	SOLVED THE CROSSWORD PUZZLE	SPOTTED MIZAR AND ALCOR	PLANE OR SATELLITE RUINED IMAGE	SPOTTED ORION'S BELT

CAN YOU SPOT THE DIFFERENCES?

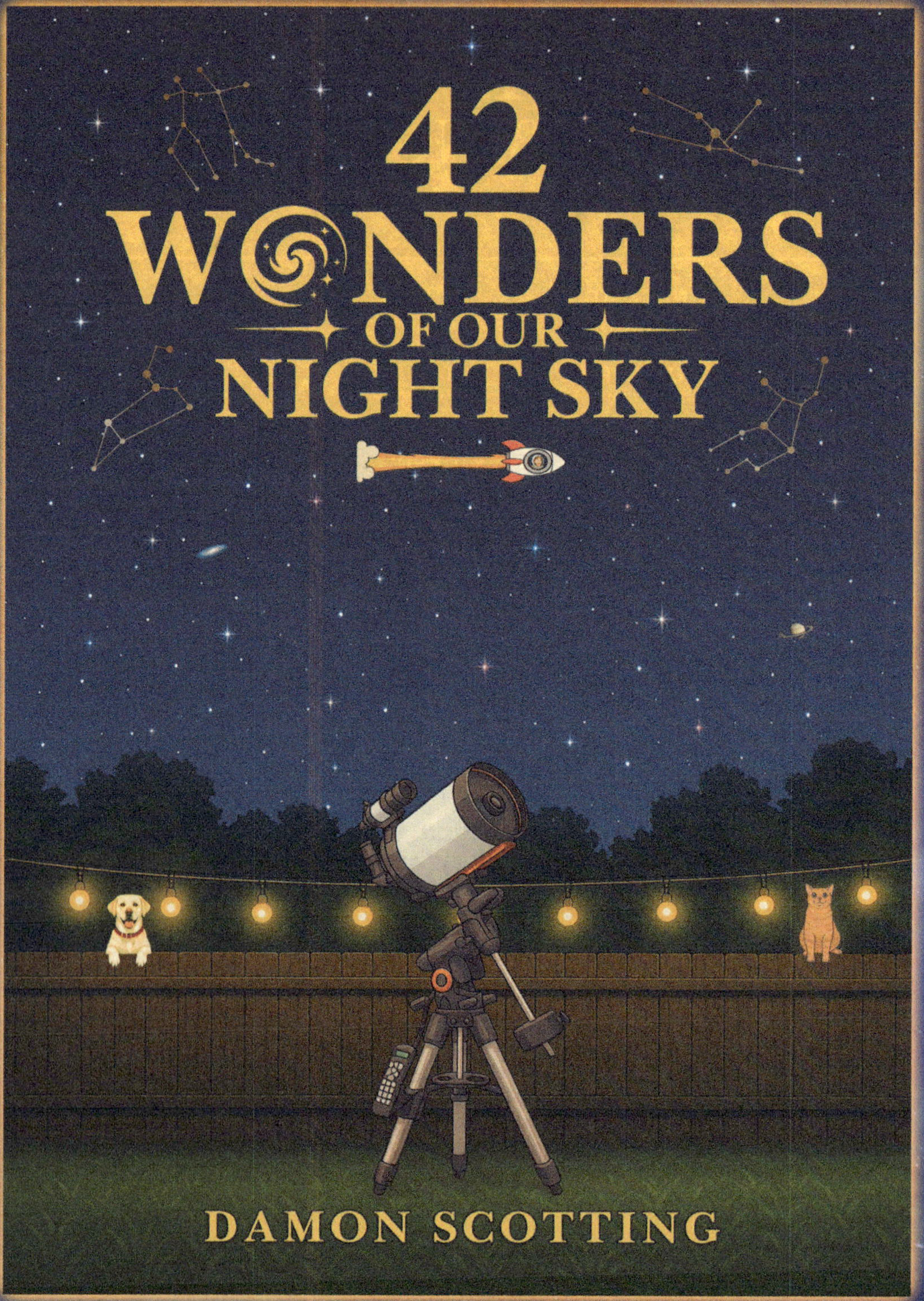

THERE ARE 10 DIFFERENCES BETWEEN THESE TWO IMAGES.

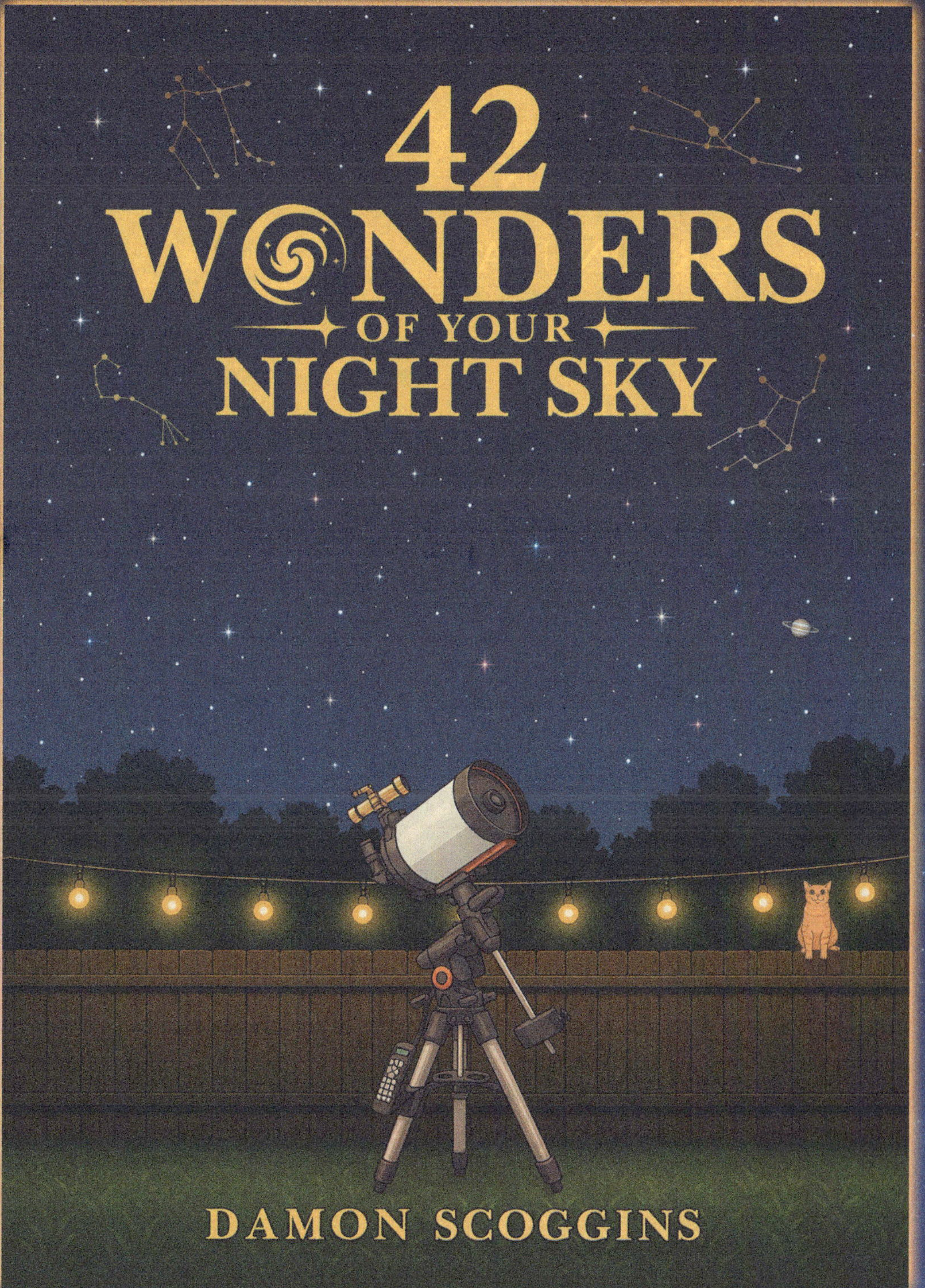

WORD JUMBLE

1. JPIRETU ______________
2. ONOM ______________
3. NEATENAN ______________
4. PERSONUVA ______________
5. NSU ______________
6. PILLOWHRO ______________
7. STEREOT ______________
8. EXLIH ______________
9. BLUEBH ______________
10. OSHEAREDH ______________
11. WORRIESFK ______________
12. ERDOWN ______________
13. ABOILER ______________
14. NEPHEWIL ______________
15. BROOMERS ______________

1. Jupiter 2. Moon 3. Antennae 4. Supernova 5. Sun 6. Whirlpool 7. Rosette 8. Helix
9. Hubble 10. Horsehead 11. Fireworks 12. Wonder 13. Albireo 14. Pinwheel 15. Sombrero

1. 

2. 

3. 

4. 

5. 

6. 

7. 

8. 

Mars

Andromeda Galaxy

Veil Nebula

Sombrero Galaxy

Whirlpool Galaxy

Crab Nebula

Pillars of Creation

Heart Nebula

1. Crab 2. Whirlpool 3. Pillars 4. Andromeda
5. Heart 6. Sombrero 7. Mars 8. Veil nebula

CAN YOU SOLVE THE PUZZLE & UNLOCK THE BONUS CONTENT?

Down

1. Hottest planet in our solar system.
2. Long blue galaxy (NGC 4631).
3. Apple core nebula (M27).
4. The "Hat" galaxy.
7. Spiral galaxy, recent supernova.
9. _______ of creation.
10. Orange & Blue Binary star.
15. M45, nicknamed the seven _______.
16. Least dense planet in our solar system.
18. Galaxy triplet, includes a Hamburger.

Across

5. Gas giant, butt of all jokes.
6. Edge-on galaxy, ranked 17th.
8. Popular winter constellation.
11. Sextuplet star system: Mizar &...
12. Eye of God nebula.
13. Swarm of stars cluster.
14. Only man-made wonder.
17. Demon star that blinks.
19. The red planet.
20. Number-one Wonder.

Solve the puzzle, then **correctly** arrange the letters in the gold boxes to reveal the password. Scan the QR code to unlock the bonus page.

Damonscotting.com/ Bonuscontent

(Prize access limited to the first 100 users to successfully enter the correct password into the web portal - Hint: Boom?)

Password

1
2
3
4
5
6
7
8
9
10
11
12
13
14
15
16
17
18
19
20

Life cycle of a star
42 Wonders edition

Star Formation

Gravity collapses cold gas and dust into a protostar, a baby star.

Star Clusters

Most stars form together, star clusters are just stellar nurseries that have slowly drifted apart as each star finds its place in our cosmos.

Main Sequence

This is the most stable period of a star's life.
Stars like our Sun will live happily for billions of years, but those that burn biggest and brightest also live for the shortest period of time.

Planetary Nebula

This is the glowing shell of gas that Sun-like stars eject near the end of their lives, lit up by the hot core that will become a white dwarf.

Neutron Star or Black Hole

The really big stars go SUPERNOVA at the end of their lives and leave behind a Neutron star, the 2nd densest object known. But the truly MASSIVE stars don't form a star, they collapse and form a literal void in space, a BLACK HOLE. We estimate that there are roughly 100,000,000 black holes in our galaxy alone.

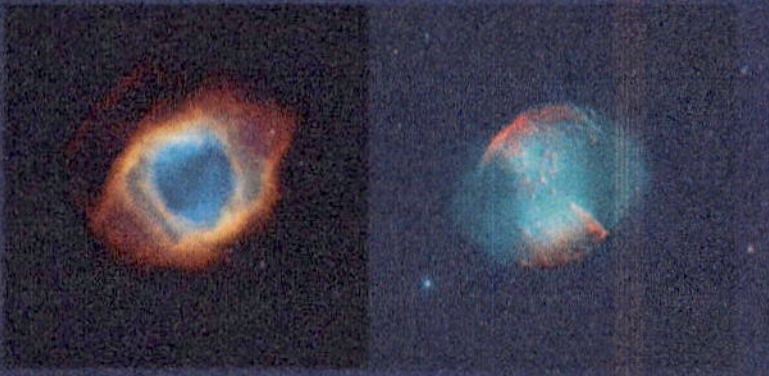

The Nebula fades & the white dwarf slowly cools & fades for billions of years.

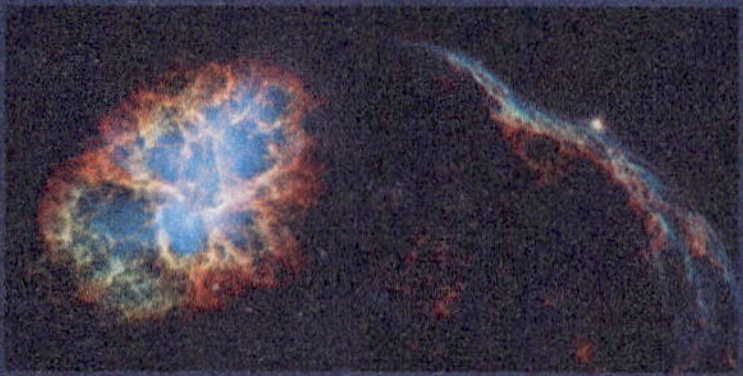

The gas & dust from dying stars is recycled into space & eventually forms new stars, planets & even life.

How BIG is it through your telescope?

This is how big these deep sky objects 'Look' compared to our Moon.

1. Helix Nebula
2. Veil Nebula
3. Pleiades
4. Andromeda Galaxy
5. Dolphin Nebula
6. Moon
7. Heart Nebula
8. Elephant Trunk Nebula
9. Orion Nebula
10. Rosette Nebula
11. Whirlpool Galaxy
12. Crab Nebula
13. Pillars of Creation
14. Sombrero Galaxy
15. Dumbbell Nebula
16. Hercules Cluster

WHAT YOU CAN IMAGE WITH A SEESTAR SMART TELESCOPE

Astrophotography is now easier than ever, and it's all thanks to the recent development of smart telescopes. These powerful, portable observatories can find deep-sky objects in our night sky and capture real images, all while you control everything from your smartphone or tablet.

They're a game changer, especially because it has always felt like you needed to invest a few thousand pounds into the hobby to take any semi-decent images. But the Seestar range of smart telescopes allows you to capture stunning, print-worthy images of our night sky for as little as $349.

There are no monthly subscription fees or essential additional accessories. The telescope is powered by its own battery, all you have to do is place it outside on a clear night, and with a few taps on your phone, it will start finding, tracking, and imaging the cosmos for you. How cool is that!

I'm a huge advocate for these smart telescopes, since they offer a cheaper gateway into the hobby than ever before. They're also incredibly intuitive and very easy to use, whether you're 6 years old or 60.

Below is a series of images I captured using each of these telescopes, along with QR codes that link directly to their respective product pages.

Specifications:

Aperture -	**30mm**
Focal length -	**150mm**
Sensor -	**IMX 662**
Resolution -	**1080P**
Optical -	**Triplet APO**
+Wide angle -	**Yes**

Seestar S30

$349

Specifications:

Aperture - 50mm
Focal length - 250mm
Sensor - IMX 462
Resolution - 1080P
Optical - Triplet APO
+Wide angle - No

Seestar S50 $499

Seestar S30 Pro $599

Specifications:

Aperture - 30mm
Focal length - 160mm
Sensor - IMX 585
Resolution - 4K
Optical - 4-Element APO
+Wide angle - Yes

What Telescope to buy?

Now that you have got a taste for the night sky it's time to find a telescope that let's you best explore it.

Shopping for astrophotography gear can be very messy. There are loads of telescopes, mounts, and cameras that look similar, cost a fortune, and can still leave you with a setup that's totally wrong for what you want to photograph. Which is why I built my very own "Astro Rig Builder" to cut through the confusion and make choosing gear feel simple, logical, and genuinely fun.

The Astro Rig Builder lets you mix and match real telescopes, mounts, and cameras and instantly see the total cost, key specs, payload safety, and field of view update live. Every selection also includes a direct product link to buy that exact item from a local retailer, so if you find "the one", you can go straight to it without hunting around online.

Personally I think the best way to use it is to just play. Build a realistic setup you could actually buy right now, or go crazy and create your dream rig, then scroll down and see real images captured with that setup (or a very similar one) so you know what results to expect.

I update this webpage every time I try out a new piece of gear, so be sure to have a play around. The website is Damonscotting.com/astro-rig-builder or you can just scan the QR code to start building your own rig for free!

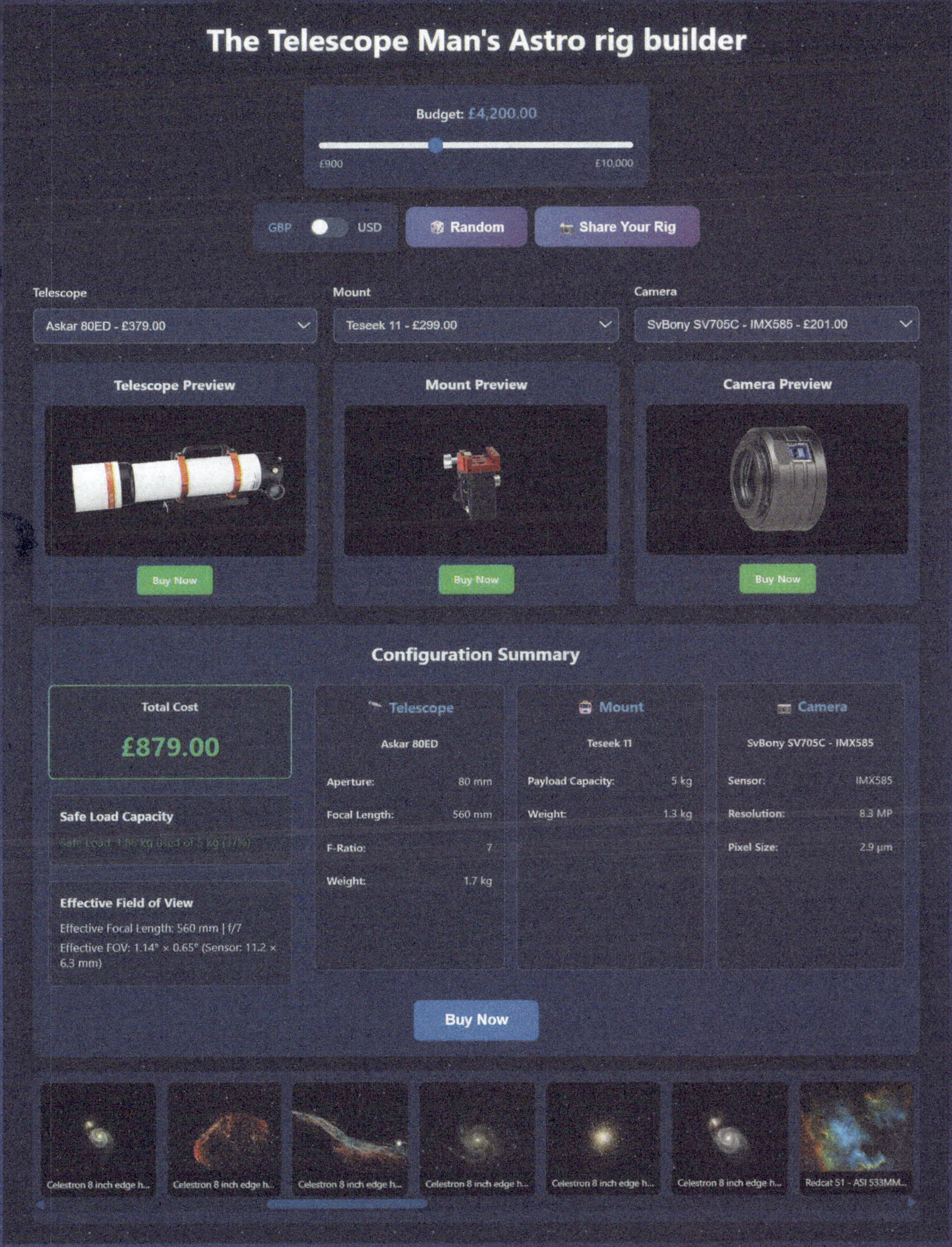
The Telescope Man's Astro rig builder
Budget: £4,200.00
£900
£10,000
GBP
USD
Random
Share Your Rig
Telescope
Askar 80ED - £379.00
Mount
Teseek 11 - £299.00
Camera
SvBony SV705C - IMX585 - £201.00
Telescope Preview
Buy Now
Mount Preview
Buy Now
Camera Preview
Buy Now
Configuration Summary
Total Cost
£879.00
Safe Load Capacity
Effective Field of View
Effective Focal Length: 560 mm | f/7
Effective FOV: 1.14° × 0.65° (Sensor: 11.2 × 6.3 mm)
Telescope
Askar 80ED
Aperture: 80 mm
Focal Length: 560 mm
F-Ratio: 7
Weight: 1.7 kg
Mount
Teseek 11
Payload Capacity: 5 kg
Weight: 1.3 kg
Camera
SvBony SV705C - IMX585
Sensor: IMX585
Resolution: 8.3 MP
Pixel Size: 2.9 µm
Buy Now
Celestron 8 inch edge h...
Celestron 8 inch edge h...
Celestron 8 inch edge h...
Celestron 8 inch edge h...
Celestron 8 inch edge h...
Celestron 8 inch edge h...
Redcat 51 - ASI 533MM...

How it was made

The 42 Wonders of Our Night Sky is a series I have been trying to create for several years now. It first started off as a Shorts-based series, where I tackled each wonder individually over the course of 42x 180-second episodes.

But the episodes fell flat; there seemingly wasn't enough of a connection between the general public and each of these night-sky objects.
So I decided instead to go all in and create a new TV-series-style format in which all 42 of the Wonders felt distinctly connected.
In the new design, each of these night-sky Wonders was judged by the same three criteria and ranked on an ultimate leaderboard to determine which was the best Wonder of our night sky.

To avoid each object still feeling abstract, I added a guided tour of each Wonder and used the insanely popular $500 Seestar smart telescope as one of the core pillars of my image comparisons.
This provided a gold-standard reference for what any of us amateurs can hope to achieve when trying to image each one of these Wonders. The result was 14 episodes of wonder-filled stargazing, 18+ months of hard work concentrated into just over three hours of stellar content!

Scan the QR Code to watch

1. I finished the full 42 Wonders list by the end of 2024. My intention was to start filming at the start of 2025, but within the first two weeks we had severe flooding and my garden was unusable.
2. The original studio setup was a bit more bare-bones compared to the final product...
3. I started trying to visualise what the ideal setup would be for this series.
4. £900 later, I had a cosy garden studio to start filming in, but it still wasn't 100%.
5. In order to add movement and a sense of professionalism, I added a motorised camera slider to incorporate a constant sense of progression in the episodes.
6. Add in some space-themed posters and pillows and the studio was looking the real deal.
7. In order to see what it looked like, I asked Hovis to be my stand-in. He was great, very professional. 10/10 "good boy".
8. Now, with a fresh taste for fame, Hovis was reluctant to give up the spotlight.
9. With the stage set, it was time to run through my lines. Fortunately, I had a bit of help from Toulouse.
10. The series utilised a three-camera rig setup. Having an actual big board with each of the Wonders and their rankings was essential to me, but designing, printing, and accounting for reflections made it a very tedious job. Overall, it was a lot of hard work, but I'm really happy with how the series has turned out. Hopefully I'll get the chance to make another one in the future.

1.
2.
3.
42 WONDERS OF OUR NIGHT SKY
4.
5.
WONDERWALL
6.
WONDERWALL
7.
42 WONDERS OF OUR NIGHT SKY
8.
RONNIE
42 WONDERS
9.
WONDERWALL
10.
THE WONDERWALL
42 WONDERS OF OUR NIGHT SKY
ANDROMEDA GALAXY
PLEIADES
SATURN
JUPITER

GLOSSARY

A

• Averted Vision: An observational technique that uses the eye's peripheral sensitivity to spot faint objects. In the book, this is a "Pro Tip" for viewing the Ring Nebula; looking slightly off-center through the telescope helps reveal details that might otherwise be invisible.

B

• Binary Star System: A system where two stars are gravitationally bound and orbit one another. Albireo is a visual binary famous for its contrasting orange and blue colors caused by temperature differences, while Algol is a complex system where three stars orbit each other, though we primarily see two interacting.

• Black Hole: An incredibly dense region of space with gravity so strong that not even light can escape. A supermassive one can be found at the center of the galaxy M87, this "freak of nature" shoots out a jet of plasma thousands of light-years long.

D

• Dark Matter: An invisible form of matter that makes up a significant portion of the universe's mass. Its presence is inferred in the Pinwheel Galaxy (M101) because the galaxy's visible stars are not massive enough to account for its rotation speed, suggesting a halo of invisible dark matter is holding it together.

E

• Eclipsing Binary: A variable star system where a fainter star passes in front of a brighter one from our perspective, blocking its light. Algol (the "Demon Star") is the classic example, dimming regularly as its "vampire" companion star eclipses the primary star.

• Elliptical Galaxy: A galaxy shaped like a massive, smooth ball of stars, lacking the spiral arms found in other types. M87 is a giant elliptical galaxy and one of the largest in the nearby universe, containing mostly old, evolved stars.

• Emission Nebula: A cloud of gas that glows with its own light, often powered by radiation from nearby stars. The North America Nebula is a prime example, glowing red due to ionized hydrogen gas.

F

• Flocculent Spiral: A type of spiral galaxy where the arms are patchy and fluffy rather than continuous and well-defined. The Pinwheel Galaxy is classified as a flocculent spiral, contrasting with "grand design" spirals that have clear, organized structures.

G

• Galactic Cannibalism: The process where a large galaxy absorbs smaller companion galaxies due to gravitational pull. The Needle Galaxy shows evidence of this violent past, with traces of absorbed galaxies visible in its halo.

• Globular Cluster: A spherical, densely packed collection of stars that orbits a galaxy. These are ancient relics, like Messier 3 and the Hercules Globular Cluster, which can be over 11 billion years old and contain stars packed as close as 0.1 light-years apart.

I

- Ionized Gas: Gas that glows because radiation from hot stars has stripped electrons from its atoms. This process creates the colors seen in astrophotography, such as the red/pink of hydrogen in the Rosette Nebula and the green/blue of oxygen in the Ring Nebula.

L

- Lenticular Galaxy: A galaxy that serves as a bridge between spiral and elliptical types, featuring a central bulge and a flat disc but no spiral arms. The Sombrero Galaxy is the most famous example, looking like a "sombrero hat" due to its prominent dust lane.
- Light Year: A unit of distance representing how far light travels in one year (about 9.46 trillion km). It is used to measure deep sky objects; for instance, the Andromeda Galaxy is 2.5 million light-years away, meaning we see it as it appeared 2.5 million years ago.

M

- Magnitude: A number that indicates the brightness of an object in the sky; the lower the number, the brighter the object. The Sun is a blinding -26.74, while the faint Antennae Galaxies are magnitude 11.2.

O

- Open Star Cluster: A group of stars that were born from the same cloud of gas and move through space together as a "cosmic family." The Pleiades and the Beehive Cluster are bright examples that can often be seen with the naked eye.

P

- Planetary Nebula: A shell of glowing gas ejected by a Sun-like star at the end of its life. Despite the name, they are not related to planets but are the "future ghosts" of stars; examples include the Dumbbell Nebula and the Helix Nebula (the "Eye of God").

S

- Spiral Galaxy: A galaxy consisting of a flat, rotating disk containing stars, gas, and dust that winds into spiral arms. The Whirlpool Galaxy was the very first object to be identified with this spiral structure.
- Starburst Galaxy: A galaxy experiencing an exceptionally high rate of star formation, often triggered by collisions with other galaxies. The Cigar Galaxy is forming stars 10 times faster than the Milky Way due to these gravitational interactions.
- Supernova Remnant: The expanding cloud of debris and shockwaves left over after a massive star explodes. The Veil Nebula is a remnant where shockwaves still travel at 1.5 million km/h, while the Crab Nebula is the remains of a star recorded exploding in 1054 AD.

V

- Variable Star: A star that changes in brightness over time. This can occur because the star physically swells and shrinks like a heartbeat, as seen with Betelgeuse, or because it is part of an eclipsing binary system like Algol.

W

- White Dwarf: The hot, dense core left behind after a star sheds its outer layers to form a planetary nebula. In the Ring Nebula, this tiny, "lonely" remnant blasts ultraviolet radiation into the surrounding gas, causing it to glow.

Acknowledgments and Image Credits

This book was created using a combination of open astronomical tools, space mission imagery, remote observatory datasets, and reference catalogues. My sincere thanks go to the scientists, engineers, and communicators whose work makes the Universe accessible to everyone.

Stellarium (stellarium.org) was used to create and verify sky maps, star-field references, object positions, and planning visuals.

Unless otherwise noted in the relevant caption or figure credit, space-based imagery and visual aids were sourced from publicly available mission archives and outreach releases, credited as follows:
Hubble Space Telescope (HST): NASA/ESA Hubble Space Telescope (via ESA/Hubble and/or STScI/HubbleSite; credits as listed with each figure). Reference: esahubble.org and stsci.edu/hst

James Webb Space Telescope (JWST): NASA, ESA, CSA, STScI, J. DePasquale (STScI), A. Pagan (STScI), A. M. Koekemoer (STScI). Reference: science.nasa.gov/mission/webb and esawebb.org

Spitzer Space Telescope: NASA/JPL-Caltech/Z. Balog (Univ. of Arizona/Univ. of Szeged). Reference: science.nasa.gov/mission/spitzer and jpl.nasa.gov/missions/spitzer-space-telescope

Algol system visual reference: Dr. Fabien Baron, Dept. of Astronomy, University of Michigan, Ann Arbor, MI 48109-1090. Reference: lsa.umich.edu/astro

Remote-observatory datasets and downloadable deep-sky imagery were sourced via Telescope Live (telescope.live), with credit to the Telescope Live platform and participating observatory hosts/operators.

Astronomical distances, magnitudes, and coordinates were referenced using SIMBAD (simbad.u-strasbg.fr) and ruuth.xyz.

Editorial tools, including Google's NotebookLM (notebooklm.google) and OpenAI (openai.com), were used for proofreading, character designs and copy editing.

My Mum, for painstakingly sticky-noting every error she found in the final draft of this book.

Adobe Photoshop (adobe.com/products/photoshop.html) was used to format and lay out this book, and PixInsight (pixinsight.com) was used to process and edit astrophotography images.

Issey, for her invaluable assistance in formatting this book in a professional manner and for keeping me on the right track. Seriously, this book would've taken a lot longer without her help.
G.O.A.T

Hovis and Toulouse, for their company over the course of the 1,000 hours this book has taken to complete!

Made in the USA
Monee, IL
03 November 2026

67379955R00079